The Laplacian Press Series
on Electrostatics

Joseph M. Crowley, Editor

I0131462

The Laplacian Press Series
on Electrostatics

Joseph M. Crowley, Editor

This series is intended to provide information on electrostatics to all levels of readers. It includes reprints of classics as well as selected original works.

ELECTROPHOTOGRAPHY AND DEVELOPMENT PHYSICS,
Revised 2nd Edition
by L. B. Schein (ISBN: 1-885540-02-7)

ELECTROSTATICS: EXPLORING, CONTROLLING AND USING STATIC
ELECTRICITY, 2nd Edition, including The Dirod ManuaL
by A. D. Moore (ISBN: 1-885540-04-3)

CONTACT AND FRICTIONAL ELECTRIFICATION
by W. R. Harper (ISBN: 1-885540-06-X)

ELECTRETS, 3RD EDITION
vol. 1, edited by G. M. Sessler (ISBN: 1-885540-07-8)
vol. 2, edited by R. Gerhard-Multhaupt (ISBN: 1-885540-09-4)

FUNDAMENTALS OF APPLIED ELECTROSTATICS
by J. M. Crowley (ISBN: 1-885540-11-6)

FUNDAMENTALS OF APPLIED ELECTROSTATICS

Joseph M. Crowley
Applied Electrostatics Research Laboratory
Department of Electrical and Computer Engineering
University of Illinois, Urbana

Δ^2 **Laplacian Press**
Morgan Hill, California

Copyright © 1999 by Electrostatic Applications. All rights reserved.

Published by Laplacian Press
A Division of Electrostatic Applications
16525 Jackson Oaks Drive
Morgan Hill, California 95037-6932, USA
Telephone: (408) 779-7774   Fax: (408) 779-3638   email: electro@electrostatic.com

Originally published by Wiley, New York © 1986
Earlier corrected reprinting by Krieger, Malabar, Florida © 1991

Publisher's Cataloging in Publication Data

Crowley, Joseph M. 1940-
    Fundamentals of Applied Electrostatics / Joseph M. Crowley
    p.   cm.
    Includes bibliographical references and index.
    ISBN 1-885540-11-6 (printed on acid-free paper)   CIP
    1. Electrostatics.   I. Title   II. Author
    (QC571.C76 1999)
    537'.2—dc20   99-067687

The use of general descriptive names, registered names, trademarks, etc. in this publication does not imply, even in the absence of a specific statement, that such names are exempt from the relevant protective laws and regulations and therefore free for general use.

Transactional Reporting Service
Authorization to photocopy items for internal or personal use, or the internal or personal use of specific clients, is granted by Electrostatic Applications, provided that the appropriate fee is paid directly to Copyright Clearance Center, 222 Rosewood Drive, Danvers MA 01923, USA.

Academic Permissions Service
Prior to photocopying items for educational classroom use, please contact the Copyright Clearance Center, Customer Service, 222 Rosewood Drive, Danvers MA 01923, USA. (508) 750-8400.

1
Printed in the United States of America

To my parents
Joseph E. and Mary V. Crowley

PREFACE

For the past twenty years I have been using electrostatics to design and analyze devices used in applications ranging from office copiers to nuclear fusion reactors. Much of this work was done on a consulting basis, and I have frequently observed that many problems arise because the electrostatics which most engineers and physicists receive in their undergraduate courses on electromagnetism is woefully inadequate, and occasionally misleading. My own understanding of the need for a fundamental approach to modern electrostatics came from two sources. One was the electrostatics lecture-demonstration of A. D. Moore, which often leaves experts in electromagnetics amazed by seemingly impossible phenomena. The other source was my colleagues in industry, especially Henry Till, who routinely showed that complicated solutions of Laplace's equation had little to do with understanding and using electrostatics.

Ignorance of electrostatics might have been acceptable when radiation and antennas were the overriding concerns in applied electromagnetism, but today there are large industries, such as office copier and computer peripheral manufacturing, which depend critically on a clear understanding of how the basic principles of electrostatics can be translated into practical devices.

To help fill this educational gap, I began several years ago to teach a course in applied electrostatics at the University of Illinois, which prerequires only the first course in electromagnetism. Although I intended it primarily to prepare my graduate students for thesis research, the course immediately attracted an undergraduate audience and is now a regularly listed elective for both undergraduate and graduate students.

This course was modeled after the physical electronics courses now taught to most undergraduate electrical engineers in which the theory provides a framework for the course and is used to explain device operation, but does not represent an end in itself. The mathematical aspects of Laplace's equation receive very little attention here, since I have found that excessive attention to mathematical complexity obscures the physical meaning of the result. Instead, a few carefully selected solutions of the electrostatic field equations are used over and over throughout the course in different contexts to describe different devices such as copy machines, smoke detectors, nonimpact printers, or the electret pickups which now account for more than half the current microphone output.

In the same spirit, the mechanical side of electrostatics is restricted to the simplest models required to illustrate the effects. Electrons are treated from a semiclassical viewpoint which makes little use of relativity, quantum mechanics, or statistical physics. Larger entities are usually considered as rigid bodies, with fluid motion and elastic deformations neglected. Although these effects must be considered in many applications, they usually appear as a modification of the basic electrostatics. In addition, their effects are already well documented in the literature.

The book follows the general plan of the course, which consists of four major parts: electrostatic fields, electrostatics of particles, electrostatics of materials, and electrostatics of circuit elements. Each chapter is composed of several sections based on individual lectures. Within each section, some basic aspect of electrostatic theory is introduced, and a device which depends on this aspect is described. The device is then analyzed using the theory, and the implications of the result for the operation of this and similar devices are discussed. A bibliography is included in each chapter so that the ideas just introduced can be explored in more detail. Because of the wide range of topics covered, it was impossible to list original articles. Instead, books and review articles were given preference, so that the reader would have access to as much further information as possible.

Although this book grew out of an upper-level university course, it should also be helpful for practicing engineers and scientists who need to understand and apply the principles of electrostatics in their daily work. These readers will naturally be most interested in their particular needs and may lack the time to work through the entire book from start to finish. To ease their task, an epilogue describes how electrostatics is applied in selected industries, sciences, and technologies, and also gives a brief guide to the most important literature in each area.

This book might have remained a set of lecture notes without the encouragement of my colleagues, especially Edward Jordan, Mac van Valkenburg, Nick Holonyak, Mangalore Pai, and Keith Watson. I owe special thanks to James Melcher for his sound advice over the years, as well as for his helpful suggestions on the manuscript and outline. Most of the typing and retyping was carried out by Janice Stephen, who always had the chapters back before I expected them. Several of my colleagues and students, especially John Chato, Darryl Hardina, Bret Jones, and Philip Krein, were kind enough to help in clarifying obscure points and correcting mistakes.

A project of this nature entails numerous evenings and weekends which I was not able to share with my family. I deeply appreciate the support of my wife, Barbara, and my sons, Joseph W., James, Kevin, Michael, and Daniel, for the last two years.

JOSEPH M. CROWLEY

Champaign, Illinois

CONTENTS

CONTENTS

NOMENCLATURE

A	area (m^2)	k	Boltzmann constant = 1.381 $\times$ 10^{-23} (J/K)
a	position (m)		
b	position (m)	L	inductance (H)
C	capacitance (F)	l	length (m)
c	position (m)	m	mass (kg)
D	diffusion coefficient (m^2/s)	N	number density (m^{-3})
D	electric displacement (C/m^2)	n	number density (m^{-3})
d	distance (m)	$\mathbf{n}$	normal vector
E	electric field (V/m)	P	polarization (C/m^2)
e	2.71828 . . .	PE	potential energy (J)
f	force density (N/m^3)	p	dipole moment (C-m)
F	force (N)	p	positive particle density (m^{-3})
G	conductance (S)	Q	lumped charge (C)
G	particle generation rate (m^3 s)$^{-1}$	q	particle charge (C)
g	gravitational acceleration $\simeq$ 9.8 (m/s^2)	R	resistance (Ω)
		R	particle recombination rate (m^{3}s)$^{-1}$
I	current (A)		
i	current (A)	r	radial coordinate (m)
$\mathbf{i}$	unit vector	s	distance (m)
J	current density (A/m^2)	T	temperature (K)
j	$\sqrt{-1}$	T	stress (Pa)
K	constant	t	time (s)
K	drag coefficient (kg/s)	U	velocity (m/s)

u	velocity (m/s)		μ	mobility (m^2/V-s)
$\mathcal{V}$	volume (m^3)		ξ	variable of integration
V	voltage (V)		π	3.14159...
v	voltage (V)		ρ	charge density (C/m^3)
W	energy		ρ_s	surface change density (C/m^2)
x	coordinate (m)		ρ_ℓ	line charge density (C/m)
Y	admittance (S)		σ	conductivity (S/m)
y	coordinate (m)		τ	torque (N-m)
Z	impedance (Ω)		Φ	potential (V)
z	coordinate (m)		ϕ	angle (rad)
α	material polarizability (C/V-m)		ψ	polar coordinate (rad)
β	reaction coefficient (units depend on mechanism)		ω	frequency (rad/s)
Γ	particle flux (m^{2}s)$^{-1}$			

Subscripts

bk breakdown value
m maximum or minimum value

γ	mass density (kg/m^3)
δ	a small quantity
ϵ	permittivity (F/m)
η	dynamic viscosity (Pa-s)
θ	angular coordinate (rad)
κ	relative dielectric constant ($\varepsilon/\varepsilon_0$)

Superscripts

+ a slightly greater value
− a slightly smaller value

PART

I

ELECTROSTATIC FIELDS

1

FUNDAMENTAL CONCEPTS

Electrostatics, as used here, involves charges in motion as well as at rest. The basic difference between electrostatics and other branches of electromagnetism is the absence of magnetic effects. The charges that move in electrostatics will of course generate magnetic fields, but in electrostatics the effects of these magnetic fields are negligible. In particular, the electric fields or voltages induced by changes in magnetic flux are much smaller than the other voltages, and the magnetic forces acting on currents are much smaller than the electrostatic forces acting on charges.

There are five fundamental quantities in electrostatics which are involved in almost all applications. Three of them, voltage, charge, and current, involve relations which are invariant; the other two, capacitance and resistance, depend on the particular materials involved. Each of these quantities may be expressed in one of two ways, depending on whether we are primarily interested in the detailed internal workings of an electrostatic device or in the gross behavior as viewed from the outside. For example, in designing a high voltage power supply, the output voltage can be calculated in terms of the terminal capacitance of the capacitors involved, and gross behavior is adequate. If we were to design the capacitors to store a given amount of charge, however, we would need to know the internal construction and the electrostatic properties of the materials involved, especially the permittivity. In this example capacitance is the gross quantity, whereas permittivity is its equivalent on a microscopic scale.

1.1 VOLTAGE AND ELECTRIC FIELD

For most engineers and scientists, the most important of the fundamental electrostatic quantities is the voltage, or electromotive force (EMF). The word force reflects the fact that voltage is related to the force applied to a charge and is derived from the electrostatic force law

$$\mathbf{F} = q\mathbf{E} \tag{1.1.1}$$

The $\mathbf{E}$ vector is thus the force experienced by a unit charge. This force is conservative, which may be expressed by

$$\oint q\mathbf{E} \cdot d\mathbf{r} = 0 \tag{1.1.2}$$

or in the equivalent differential form as

$$\nabla \times \mathbf{E} = 0 \tag{1.1.3}$$

The definition of the voltage follows from the application of the line integral around a path that includes the terminals of a device, such as the parallel plate capacitor sketched in Figure 1.1.1. Following the path indicated, which passes across the terminals, through the conductors, and across the interior gap, gives the line integral in the form

$$\oint \mathbf{E} \cdot d\mathbf{r} = \int_{-\text{terminal}}^{+\text{terminal}} \mathbf{E} \cdot d\mathbf{r} + \int_{\text{wires}} \mathbf{E} \cdot d\mathbf{r} + \int_{\text{gap}} \mathbf{E} \cdot d\mathbf{r} = 0 \tag{1.1.4}$$

The integral inside the wires vanishes because the electric field inside good conductors is very small, so only two line integrals are left to evaluate. Inside the capacitor, the electric field $\mathbf{E}$ is uniform, so the internal integral becomes

$$\int_{\text{gap}} \mathbf{E} \cdot d\mathbf{r} = Ed \tag{1.1.5}$$

The $\mathbf{E}$ field between the terminals is difficult to describe since it does not have the simple uniformity of the internal field. In many applications, however, a detailed knowledge of this field is not needed, and the integrated field (or voltage) suffices. The voltage is defined in terms of the first integral in Eq. (1.1.4) as

$$v = -\int_{-\text{terminals}}^{+\text{terminal}} \mathbf{E} \cdot d\mathbf{r} \tag{1.1.6}$$

Thus, in the example, the terminal voltage is given by

$$v = Ed \tag{1.1.7}$$

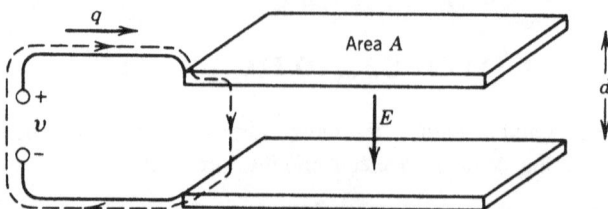

FIGURE 1.1.1. A capacitor with terminals.

The boundary condition for the electric field comes directly from the integral law applied to a surface, as shown in Figure 1.1.2. Evaluation of Eq. (1.1.2) along the path shown in the figure gives

$$(\mathbf{E}_a)_{\text{tan}} = (\mathbf{E}_b)_{\text{tan}} \tag{1.1.8}$$

Only the tangential component of the field appears in the boundary condition because the vertical legs of the path are vanishingly small.

The voltage or electric potential can also be expressed in terms of the **E** field by a differential relation which is equivalent to the line integral given before, as

$$\mathbf{E} = -\nabla\Phi \tag{1.1.9}$$

The quantity Φ is used for the voltage at any point, whereas v is usually reserved for the potential difference between two terminals.

Since the voltage and the **E** field both refer to the same physical effect, namely, the force on a charge, they are therefore closely related in any application. The relation involves the relative geometry of the terminals and the internal field distribution, which is expressed in terms of some characteristic length, such as d, the separation between plates in the example.

In applying electrostatics, we often find an extremely wide range for the fundamental variables. This is true for both the voltage and the electric field. Since voltage is usually a more familiar concept, the voltage levels associated with various effects and applications are outlined in Table 1.1.1. There are several key landmarks along this table which should be kept in mind in designing devices based on electrostatics. The most important is 300 V, where air breakdown first becomes possible. Below this voltage, components may usually be placed as close as desired, down to a few microns apart, without the need for special insulating techniques. Above this voltage, corona and sparking must be considered in the design, which usually increases both cost and size. As the voltage is raised even more, breakdown becomes progressively more difficult to prevent. At the extreme upper voltages found in thunderclouds, even separations of miles are not enough to prevent lightning breakdown.

A similar table can be prepared for the electric field **E** showing typical magnitudes in different applications (Table 1.1.2). As with voltage, the key landmark here is the minimum field at which air breakdown is possible, since design of devices is always more difficult above this point. In practice, many applications involve electric fields slightly below this level, since the electrostatic effects are then as strong as possible without the risk of breakdown.

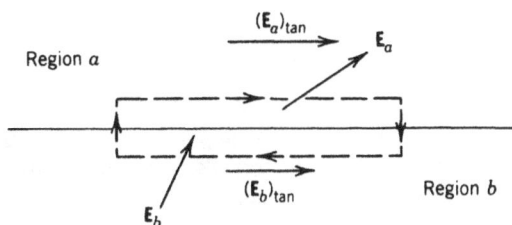

FIGURE 1.1.2. Electric fields at a boundary.

TABLE 1.1.1.   Typical Voltages

Voltage	Example
10^{-6}	Radio antenna signal output
10^{-1}	Biological cell
	$p-n$ junction
	Contact potential
1	Electrochemistry, batteries
10	Solid state devices become expensive above here
10^2	Household wiring
300	Lowest voltage for air breakdown
10^4	Power plant generators
	Ionizers
	TV sets
10^5	Power transmission
	X-ray machines
10^6	Upper limit of power transmission
	Earth-to-ionosphere voltage
	Nuclear particle accelerators
10^8	Thunderclouds

TABLE 1.1.2.   Typical Electric Fields

E Field (V/m)	Example
10^2	Fair weather atmosphere
10^5	$p-n$ junctions
	Thunderstorms
	Power transmission lines
3×10^6	Air breakdown possible
10^7	Oils, solids break down
10^8	Field emission of electrons from surfaces

It should be pointed out that there are usually two criteria for breakdown, one based on voltage and one on electric field. As a rule, both limits must be exceeded before breakdown. Since the voltage and field are related by a characteristic length, there is a critical size which separates voltage controlled breakdown from field controlled breakdown. This size is approximately

$$d = \frac{V_{bk}}{E_{bk}} = \frac{300}{3 \times 10^6} = 0.1 \text{ mm} \qquad (1.1.10)$$

Below this size, the fields may be large, but breakdown will not occur until the voltage exceeds approximately 300 V. For larger gaps, the voltage may be large, but the field must exceed 3 MV/m for breakdown. These different breakdowns are discussed in more detail in later chapters.

1.2 CHARGE AND ELECTRIC DISPLACEMENT

Along with voltage, charge is one of the two electrical variables which must be considered in any application of electrostatics. Just as the terminal voltage is related to an internal electric field, there is a field related to charge which is more useful if the configuration is complex. The appropriate field quantity is the electric displacement vector **D**, which is usually defined in terms of Gauss' law

$$\oint \mathbf{D} \cdot d\mathbf{A} = q \qquad (1.2.1)$$

The **D** vector has the dimensions of charge per unit area. Its application often takes the form of a simple integral when the system under consideration has a high degree of symmetry, such as the charged sphere shown in Figure 1.2.1. If the charge is distributed uniformly around the sphere, then the **D** field outside will have a constant magnitude and a direction which points in the radial direction at every point. Evaluation of the integral is straightforward in such a case, giving the result

$$D = \frac{q}{4\pi r^2} \qquad (1.2.2)$$

Note that the magnitude of the **D** vector is just the charge on the sphere divided by the surface area, so that the **D** vector is, in a sense, the normalized charge.

Another basic **D**-field integral occurs at the interface between two regions and leads to the basic boundary condition for **D**. The volume is usually taken as cylindrical, with the sides of the cylinder perpendicular to the interface and each of the two ends in different regions, as shown in Figure 1.2.2. For a cylinder, the contributions to the area integral in Gauss' law [Eq. (1.2.1)] can be divided into those from the ends of the cylinder and those from the side. At the ends the vertical **D** field is normal to the surface, so only the vertical components D_a and D_b make a contribution to the integral. On the side only the horizontal component of **D** can contribute to the integral. In deriving the boundary conditions we usually assume that the cylinder is very short, which implies that the side area integral becomes

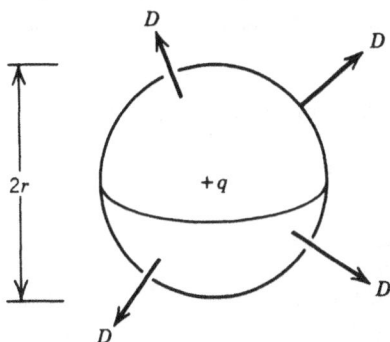

FIGURE 1.2.1. D field around a charged sphere.

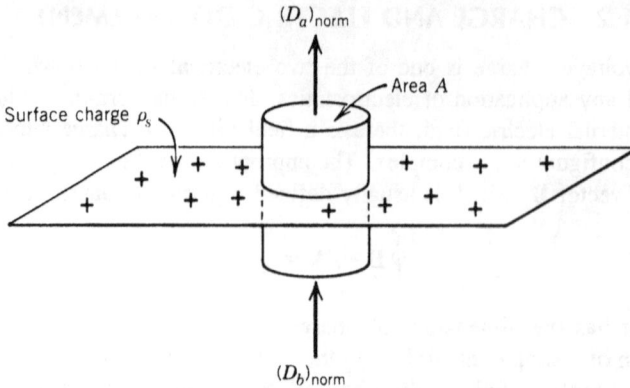

FIGURE 1.2.2. Boundary condition for D.

very small compared to the integrals over the ends. In this case the side integrals can be completely neglected, and Gauss' law becomes

$$(D_a)_{\text{norm}} - (D_b)_{\text{norm}} = \frac{q}{A} \equiv \rho_s$$

The difference in the two **D** fields is given by the charge/area ratio, as before. This quantity, the surface charge, occurs frequently in practice, since the charge on conductors tends to accumulate at the surface.

When the shape of the surface is not simple, or the charge is not uniformly distributed, it is often easier to use the differential form of Gauss' law

$$\nabla \cdot \mathbf{D} = \rho \qquad (1.2.4)$$

The charge in this form of the equation is given by the quantity ρ, which is charge/volume, or charge density. Thus the charge may be given in one of three forms, depending on the particular device or application. The fundamental expression, of course, is the charge itself, which is denoted by q. Some typical values of charge are given in Table 1.2.1. As with most of the quantities of electrostatics, the values of charge encountered in practice span an unusually wide range of values (25 orders of magnitude in the table). Usually, however, the practical effect is not related to the total amount of charge as much as to one of the charge densi-

TABLE 1.2.1. Some Typical Values of Charge

Charge (C)	Example
10^{-18}	Ions, electrons
10^{-15}	Aerosol droplets, ink jet printers
10^{-12}	Small capacitor in electronic circuit
10^{-6}	Person on synthetic carpet
10^{-3}	Large capacitor in power circuit
10^{7}	Whole earth

ties ρ_s or ρ. Electric breakdown, for instance, tends to occur in air whenever the surface charge density exceeds 25 $\mu C/m^2$, whether the charge resides on an ink drop, a person, or the earth. Thus the wide range of charges in the table reflects mostly the range in size of the object involved.

1.3 MATERIAL PROPERTIES
(Capacitance and Permittivity)

A basic concern in electrostatics is the relation between the two variables defined so far, the voltage (or **E** field) and charge (or **D** field). There is no fundamental physical law which gives this relation. Instead, the relation must be determined for each particular material, and in practice a wide variety of relations is found. This variety in material properties, which is seldom dealt with in classical electrostatics, is responsible for the great number of devices and applications in which electrostatics plays a role.

This relation is usually determined by experiments in which the voltage is changed and the charge flow measured. There are two basic types of results. In some systems, like ions or electrets, charge is present independent of the voltage. In other systems, such as capacitors, the amount of charge depends primarily on the voltage applied, although it may also depend on the history of the voltage application. The reasons for these various relations, as well as their applications, are covered in detail in later chapters. For now, it is only necessary to see how they can be used to relate the terminal variables v, q to the field variables **E, D**.

The simplest case is the parallel plate capacitor, shown in Figure 1.1.1. Usually measurements of charge and voltage show a linear relation of the form

$$q = Cv \tag{1.3.1}$$

The slope of the straight line C is called the capacitance. It has the units of coulombs per volt, or farads (F). Some typical values are shown in Table 1.3.1. Again, there is a very wide range of values encountered in practice. Most of this range, however, is due to the size variation rather than to a difference in the material properties. For this reason, it is often useful to separate the geometry from the material properties by using a field quantity related to capacitance.

In the preceeding sections, we saw that q and v both have counterparts in field vectors, which are related by the geometry of the device and the charge distribution. In a parallel plate capacitor, the symmetry ensures that the charge and fields

TABLE 1.3.1. Typical Capacitance Values

Capacitance (F)	Example
10^{-16}	Aerosol droplet
10^{-12}	Small capacitor in electronic circuit
10^{-3}	Large filter capacitor
10^2	Earth

are uniform, so the relations can be obtained from the integral equations. For the
E field the line integral becomes

$$v = Ed \tag{1.3.2}$$

whereas the charge on the upper plate (the one connected to the + terminal) is
given by Gauss' law as

$$q = DA \tag{1.3.3}$$

Substituting these expressions for charge and voltage into the experimental terminal relation gives a relation between the D and E fields of the form

$$D = \left(\frac{Cd}{A}\right)E \equiv \epsilon E \tag{1.3.4}$$

The quantity ϵ is called the permittivity of the material. It is essentially a capacitance normalized to the geometry of the device so that it has the same value for a
given material, regardless of its size or shape. Like the D and E vectors, it is
much more useful in designing or modeling a system when the shape is not simple
or symmetrical.

It should be stressed that permittivity is not a generally useful quantity. It is
only defined for materials which have a linear relation between D and E, and even
then it does not tell the whole story unless the charge vanishes when the electric
field does. There are many useful materials, mostly solids, which do not satisfy
these restrictions and can not be described in terms of a permittivity. At the same
time, many common materials, including most gases and liquids, have a linear
capacitance relation, and it is for these materials that permittivity is most useful.

Permittivity has units of farads per meter, but it is rarely given in these units.
Instead, all materials are usually referred to a vacuum, which has a permittivity of

$$\epsilon = 8.854 \times 10^{-12} \text{ F/m}$$

The normalized permittivity, or dielectric constant, is defined by

$$\kappa = \frac{\epsilon}{\epsilon_0}$$

It should be kept in mind when doing calculations that the equations (in SI units)
will be incorrect unless the permittivity is entered in units of farads per meter, so
all dielectric constants must be multiplied by ϵ_0 first.

Some typical values of the dielectric constant are given in Table 1.3.2. Most
gases have dielectric constants close to that of vacuum, whereas most liquid and
solid insulators are in the range of 2.5 to 8. Conducting liquids can be much
higher, with water among the highest of the liquids at 80. Nominal values for
solids can be quite high, but, since many solids also display nonlinear material
relations, the equivalent linear parameter (permittivity or dielectric constant) may
not be valid in a particular application.

TABLE 1.3.2. Typical Dielectric Constants

Dielectric Constant	Material
1	Vacuum, most gases
2.2	Insulating oils
2.5	polystyrene
3.4	poly(methyl methacrylate) (Plexiglas)
7.0	Porcelain
80	Water
1200	Barium titanate

1.4 CHARGE CONSERVATION AND CURRENT

The charge previously described is not restricted to a particular location; it is often free to move about in a device. In fact, many devices and applications depend on this charge motion for their operation. To describe the behavior of such devices successfully, it is necessary to have some way to keep track of the location of the charges as time progresses. The basic method for locating the charges is the law of conservation of charge, which can be explained with reference to some volume of space which contains charges, as shown in Figure 1.4.1.

The charge inside the volume can be increased by bringing in charges from outside. This flow of charges is called a charge current, or simply a current. The current is defined by the amount of charge it brings in a given time, or

$$i = \frac{dq}{dt} \tag{1.4.1}$$

The current need not be concentrated at a single place on the surface of the volume; it could result from several wires leading to the inside, in which case the current i would be the sum of all of the current contributions. Wires which carry charge out of the volume are represented by negative currents. In this case, the net current would be the algebraic sum of all of the contributions, or

$$i = \sum_k i_k \tag{1.4.2}$$

If there are many paths for charge flow, it may be convenient to think in terms of a current density, or a current distributed over the entire surface of the volume. In this case, the net current is usually written as

$$i = -\oint \mathbf{J} \cdot d\mathbf{A} \tag{1.4.3}$$

The new quantity $\mathbf{J}$ has the units of amperes per square meter, and represents the contribution to the total current which penetrates a unit area of the surface. The

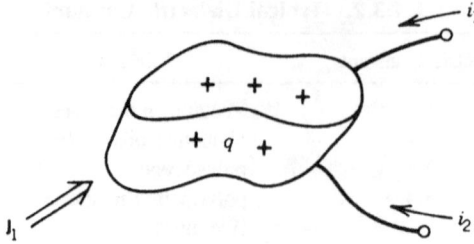

FIGURE 1.4.1. A volume containing charge.

minus sign appears in the definition because the normal to the surface is always taken to be positive in the outward direction, whereas the current is usually considered positive when it increases the charge within the volume.

At boundaries or interfaces, charge conservation takes on a special form, corresponding to the abrupt change in properties at that point. If a cylindrical volume is drawn around a section of the boundary, as shown in Figure 1.4.2, the law of charge conservation can be written as

$$(J_a)_{\text{norm}} - (J_b)_{\text{norm}} = -\frac{d\rho_s}{dt} \tag{1.4.5}$$

Since only the local portion of the boundary is involved here, the surface charge density ρ_s represents the charge inside the volume.

$\mathbf{J}$ is a field, or local quantity, which is more suitable for describing the internal workings of an electrostatic device. The conservation of charge relation can be transformed into a differential form by recalling the relation between surface and volume integrals,

$$\oint \mathbf{J} \cdot d\mathbf{S} = \iiint (\nabla \cdot \mathbf{J})\, dV \tag{1.4.6}$$

and the relation between charge and charge density

$$q = \iiint \rho\, dV \tag{1.4.7}$$

Using these two expressions gives the differential form of charge conservation as

$$\frac{d\rho_s}{dt} + \nabla \cdot \mathbf{J} = 0 \tag{1.4.8}$$

The currents which flow in electrostatic applications have a wide range of magnitudes, just as with the other electrostatic variables. Some typical values are shown in Table 1.4.1. For safety's sake, anyone who works in this field should be aware of the current levels associated with personal injury or death. Electric current can first be felt on unbroken skin near the 1-mA level, and can cause death by disrupting the heart rhythm near 100 mA.

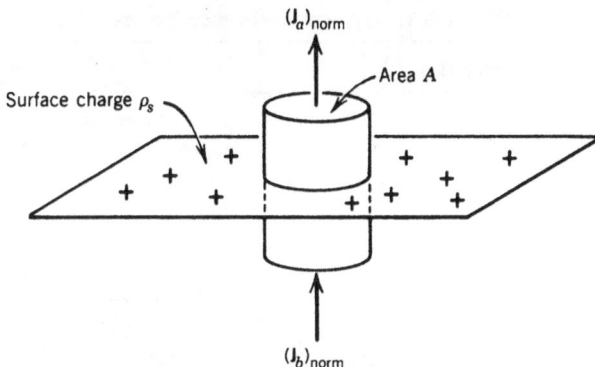

FIGURE 1.4.2.   Charge conservation at an interface.

TABLE 1.4.1.   Typical Current Magnitudes

Current (A)	Example
10^{-6}	Radio signal
10^{-3}	Threshold of shock
10^{-2}	Transistor circuits
10^{-1}	Heart fibrillation possible
10^{3}	Total earth current
10^{6}	Lightning peak

1.5   RESISTANCE AND CONDUCTIVITY

When a current flows into a device like a capacitor, the charge builds up continuously, and the voltage increases with it. In some devices, called resistors, a different pattern results. The voltage reaches a constant value and remains at that level as long as the current persists. Most commonly, the voltage is linearly proportional to the current, as described by Ohm's law,

$$v = Ri \tag{1.5.1}$$

The quantity R is the resistance of the device. It should be kept in mind that resistance is really a linear concept, since it is the coefficient in Ohm's law. Especially in highly insulating materials, where the resistance is high, it is unusual to find a linear relation between voltage and current, and the idea of resistance can often lead to erroneous conclusions if it is used in a nonlinear context. Some typical resistance values encountered in practice are listed in Table 1.5.1.

As with all the other electrostatic quantities, it is often more convenient to use a field quantity related to the resistance observed at the terminals. For example, if the device has the parallel plate geometry shown in Figure 1.5.1, then $v = Ed$,

TABLE 1.5.1. Typical Resistance Values

Resistance (Ω)	Example
10^3	Bipolar transistor circuits
10^6	Instrumentation
10^9	FET circuits
10^{14}	Electrometers

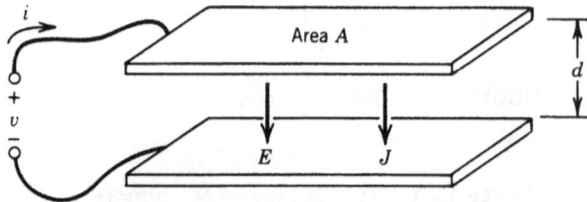

FIGURE 1.5.1. A parallel plate resistor.

$i = JA$, and Ohm's law becomes

$$J = \left(\frac{d}{AR}\right)E \equiv \sigma E$$

The quantity σ is the conductivity of the material, which is the field quantity related to the terminal quantity of resistance. The field quantity is occasionally expressed as resistivity (the reciprocal of conductivity), but conductivity is a far more convenient form when relating the measured values to the various conduction mechanisms which can arise in fluids and solids. Again, as with resistance, conductivity is basically a linear concept, and it should be used cautiously in nonlinear materials, which often are associated with low conductivity values.

Typical conductivity values for materials often encountered are listed in Table 1.5.2. Especially at the lower conductivity values, these should only be taken as rough estimates. Air, for instance, has a very low conductivity, but it easily breaks down by means of arcs or sparks when it carries much more current than expected on the basis of its low conductivity. This is a reflection of the highly nonlinear behavior associated with poor conductors.

TABLE 1.5.2. Typical Conductivity Values

Conductivity (S/m)	Example
10^{-17}	Polystyrene
10^{-12}	Glass
10^{-9}	Oils
10^{-6}	Distilled water
1	Sea water
10^7	Metals

BIBLIOGRAPHY

Fano, R. M., L. J. Chu, and R. B. Adler, *Electromagnetic Fields, Energy, and Forces*, Wiley, New York, 1960.

Hayt, W. H., Jr., *Engineering Electromagnetics*, McGraw-Hill, New York, 1967.

International Telephone and Telegraph Corporation, *Reference Data for Radio Engineers*, International Telephone and Telegraph Corp., New York, 1956.

Landau, L. D., and E. M. Lifshitz, *Electrodynamics of Continuous Media*, Addison-Wesley, Reading, MA, 1960.

Panofsky, W. K., and M. Phillips, *Classical Electricity and Magnetism*, Addison-Wesley, Reading, MA, 1965.

Plonus, M. A., *Applied Electromagnetics*, McGraw-Hill, New York, 1978.

Stratton, J. E., *Electromagnetic Theory*, McGraw-Hill, New York, 1941.

Zahn, M., *Electromagnetic Field Theory*, Wiley, New York, 1979.

PROBLEMS

PROBLEM 1   (INSTRUMENTS)

An electrostatic flowmeter is often used to monitor the breathing of patients with respiratory difficulties. It consists of three porous electrodes mounted perpendicular to the air flow. The two outer electrodes are rigid, while the middle one can deflect in response to the air movement. The electrodes, which remain parallel to each other during operation, have an area A (= 10 cm^2) and an initial separation d (= 1 mm). The center electrode is grounded, whereas the other two are driven by identical sinusoidal voltages. Find the difference in current to the two electrodes $i_{diff} = i_1 - i_2$ if the exciting voltage has an amplitude V_0 (= 10 V) and a frequency ω (= $2\pi \times 10^3$ rad/s).

PROBLEM 2   (INSTRUMENTS)

The basic charging mechanism in a van der Graaf generator is the production of charge at a low voltage and transport to a high voltage electrode. A simple model consists of the two capacitors shown in Figure 1.P.2.

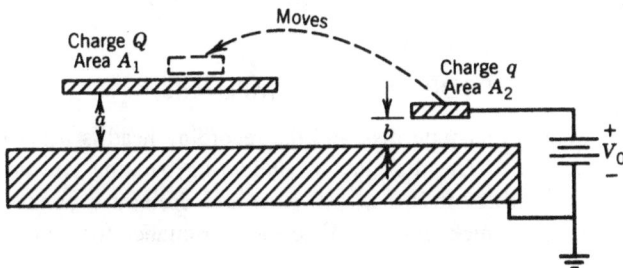

FIGURE 1.P.2.   A model van der Graff generator.

a. Find the voltage on the large capacitor, and the charge q on the small one.

b. The battery is disconnected, and the small capacitor plate is placed on top of the large one, as shown. Where on the electrode surfaces are the charges? What is the voltage between the capacitor plate and ground?

c. The small plate is lifted from the large one. What charge is left on it? What is the voltage on the large capacitor?

d. In practice, this process is carried out continuously on a moving belt of width w, which carries an area

$$dA_2 = wu\,dt$$

to the upper plate in a time dt. Typical values, using frictional (or static) charging are

$$A_1 = 0.1 \text{ m}^2, \quad w = 0.1 \text{ m}, \quad u = 1 \text{ m/s}, \quad a = 0.1 \text{ m}, \quad b = 10^{-6} \text{ m}$$

Find the voltage on the larger capacitor as a function of time if $Q = 0$ at $t = 0$.

PROBLEM 3   (HOUSEHOLD)

The CED videodisc player is based on capacitance changes which occur as the disc surface passes a stylus head, as shown in Figure 1.P.3. The video signal is related to the height of the conducting layer $x(t)$ and is decoded by a circuit which responds to the capacitance between the disc and stylus. Use a parallel plate approximation to find this capacitance, and estimate its value if the dimensions are

$$d \simeq 10 \ \mu\text{m}, \quad 0.1d < x < 0.9d, \quad A \simeq 10^{-10} \text{ m}^2$$

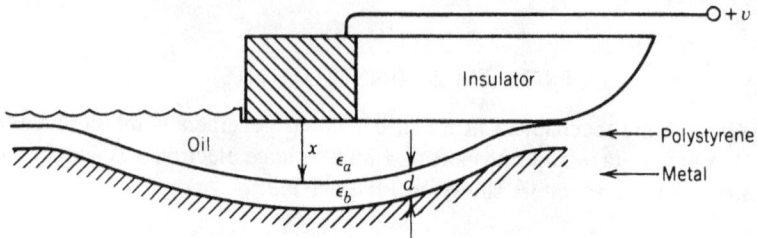

FIGURE 1.P.3.   A videodisc player pickup.

PROBLEM 4   (COMPUTERS)

The air gap between a magnetic disk and the recording head is usually measured by a capacitance probe. The disk has a conducting backing and an insulating magnetic coating of thickness d $(= 40 \ \mu\text{m})$ and dielectric constant κ $(= 5)$ while the air gap has a thickness $x(t)$. Find the capacitance for a probe of area A $(= 10^{-10} \text{ m}^2)$.

PROBLEM 5   (INDUSTRY)

During the operation of an electrostatic precipitator, an insulating layer of powder builds up on the grounded collecting electrodes, while continued precipitation carries a current of density J_0 (= 10^{-4} A/m^2) to the top of the layer. If the layer has an ohmic conductivity σ (= 10^{-12} S/m), find the voltage across the layer and the electric field inside the layer as a function of layer thickness x. Is breakdown possible in the layer?

2

ELECTRIC FIELDS
WITH
KNOWN VOLTAGES

Virtually all applications of electrostatics require knowledge of the electric fields, which are described by solutions of Poisson's equation. This equation has been under intense scrutiny for more than a century, and many useful solutions have been tabulated (some in Appendix A). In practical work these solutions must usually be adapted to the problem at hand by simplification of the geometry or by specification of voltage or charge in selected regions.

The simplest conditions are those in which the voltage is specified at electrodes and no charge exists between those electrodes. Poisson's equation then reduces to the simpler equation of Laplace,

$$\nabla^2 \Phi = 0$$

When voltage is specified over a given electrode surface, the potential $\Phi(x, y, z)$ should also be constant over the same surface to allow easy matching of the boundary condition. Since there are only a few solutions which give constant potential on a simple surface, we usually try to fit these simple shapes into the practical problem, either by design (if we are constructing the device) or by approximation (if we must deal with an existing situation). This modeling process is illustrated in the following pages for the five simple shapes which arise most often in practice. They are parallel plates, coaxial cylinders, concentric spheres, a cylinder in a uniform field, and a sphere in a uniform field.

2.1   ELECTRIC FIELDS IN TWO FLAT LAYERS
(High Voltage Bushing Design)

Summary

The simplest electrostatic problems lack space charge and have specific potentials on the boundaries. Even in rectangular geometry, these basic problems can have surprising results, as illustrated in an example of bushing design. Solution for the breakdown strength of a bushing shows that adding a better insulator to a device often causes it to break down at a lower voltage.

Theory

The simplest electrostatic situations are those in which the potential is specified at the boundaries and there is no free charge in the volume. The absence of free charge means that the potential distribution can be given by Laplace's equation,

$$\nabla^2 \Phi = 0$$

whereas boundary conditions are given directly in terms of the applied voltage.

A simple solution of this equation, which occurs over and over in practice, concerns two layers of different material placed between electrodes, as shown in Figure 2.1.1.

The electrodes and interface are all horizontal, which suggests that the potential varies only in the vertical (x) direction. The solution of Laplace's equation which has only x variation is

$$\Phi_a = K_{1a}x + K_{2a} \tag{2.1.1}$$

in the layer above the interface and

$$\Phi_b = K_{1b}x + K_{2b} \tag{2.1.2}$$

in the layer below. The electric fields are

$$\mathbf{E}_a = -\nabla\Phi_a = -K_{1a}\mathbf{i}_x \tag{2.1.3}$$

$$\mathbf{E}_b = -\nabla\Phi_b = -K_{1b}\mathbf{i}_x \tag{2.1.4}$$

Thus the electric fields in the layers above and below the interface will be uniform.

One boundary condition at the interface requires that the potential on either side be continuous,

$$\Phi_a(x = 0) = \Phi_b(x = 0)$$

or

$$K_{2a} = K_{2b}$$

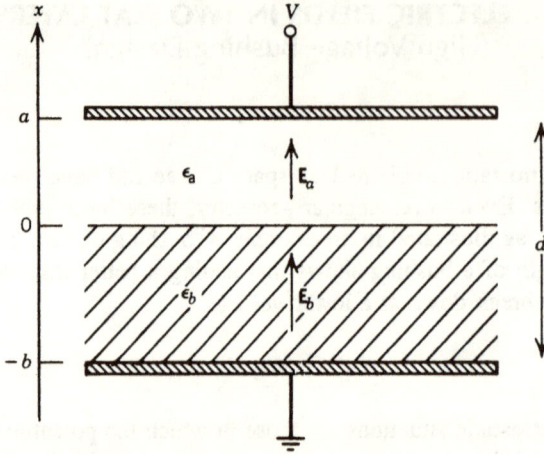

FIGURE 2.1.1. Bushing with porcelain and air.

The other condition comes from Gauss' law

$$D_a(x = 0) - D_b(x = 0) = \rho_s \tag{2.1.5}$$

In many electrostatic applications the surface charge ρ_s is an important source of electric fields and must be determined to analyze the device correctly. When dealing with ac fields or conducting materials, however, this surface charge can often be neglected, so that

$$\rho_s = 0 \tag{2.1.6}$$

(Several examples of applications in which surface charge is not negligible are given in Chap. 3). With this assumption, the boundary condition gives

$$\epsilon_a E_a = \epsilon_b E_b \tag{2.1.7}$$

or

$$\epsilon_a K_{1a} = \epsilon_b K_{1b} \tag{2.1.8}$$

using the permittivity relation.

Two additional boundary conditions come from the potentials on the upper and lower electrodes. At the upper electrode

$$\Phi_a(x = a) = K_{1a}a + K_{2a} = V_0 \tag{2.1.9}$$

while at the lower electrode

$$\Phi_b(x = -b) = -K_{1b}b + K_{2b} = 0 \tag{2.1.10}$$

Using the four boundary conditions gives the electric fields in the two layers as

$$E_a = \frac{-(\epsilon_b/\epsilon_a)(V_0/d)}{1 + [(\epsilon_b/\epsilon_a) - 1](a/d)} \tag{2.1.11}$$

$$E_b = \frac{-(V_0/d)}{1 + [(\epsilon_b/\epsilon_a) - 1](a/d)}$$ (2.1.12)

where $d = a + b$, the total width of the gap.

Example: Flat Bushings

A common but important example of this kind of problem is the design of bushings to bring high voltage ac lines into buildings, transformers, and the like. In the simplest case, the lead-in will be a flat conductor separated from the grounded container by a constant distance d. If the breakdown field strength of the bushing material is E_{bk}, then the bushing must have a minimum thickness of

$$d_m = \frac{V_p}{E_{bk}}$$ (2.1.13)

to withstand a peak voltage of V_p. Often, this thickness is too large for the application. For instance, if the bushing is insulated with air, which has a breakdown strength of 3 MV/m, then the gap must be at least 16.3 cm if a 345-kV line (with a peak voltage of 488 kV) is involved.

In this situation, an insulator such as porcelain (with a breakdown strength of 6 MV/m) might be chosen. This would reduce the required spacing to 8.1 cm and would also offer mechanical support for the conductor. Although completely filling the bushing space with porcelain would increase the voltage rating, the performance can be much worse if the space is only partially filled and some air gap is left. The reasons for this, and the actual voltage rating for a partially filled gap, follow from the basic field solution just discussed.

Application of Theory

With air ($\epsilon = \epsilon_0$) as the upper layer, the field there is given from Eq. (2.1.11) as

$$E_a = \frac{-\kappa}{1 + (\kappa - 1)(a/d)} \frac{V_p}{d}$$ (2.1.14)

where κ is the dielectric constant of the insulator $\kappa = \epsilon/\epsilon_0$. Note that the electric field in the air is *larger* than it was before we "insulated" the gap. Figure 2.1.2 shows the increase in the electric field when the insulating layer is inserted.

Discussion

Because the field in the air is larger, breakdown now occurs at a lower voltage. Since this is the exact opposite of what was intended by the use of insulation, it is important to understand what has happened here. The key idea is the uniformity of $D(= \epsilon E)$. It is uniform across the gap (in the absence of surface charge), so wherever ϵ is smaller, the E field must be larger. In effect, materials with high dielectric constants screen out the electric fields, leaving most of the voltage drop

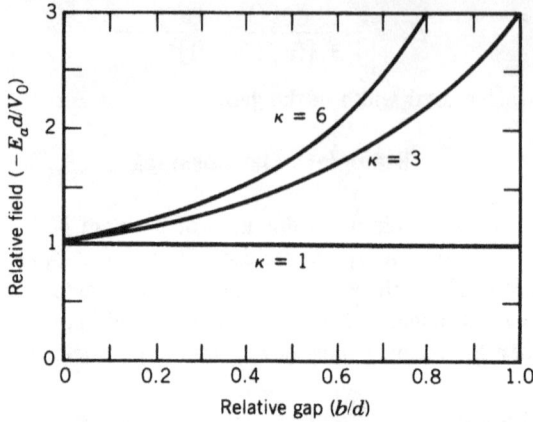

FIGURE 2.1.2.   Electric field in a partially filled air gap.

across the air. The porcelain has a higher breakdown strength, but it also has a higher dielectric constant than air, so it raises the electric field strength in the surrounding air. This shifts the burden of breakdown strength to the air, which is less able to withstand it. As a result, the bushing will break down at a lower voltage than before. When the air breaks down, it becomes a good conductor of electricity, so that the entire voltage now appears across the porcelain. Unless the air gap is very small, this increases the E field in the porcelain to a value greater than its breakdown strength, and the remainder of the bushing will be breached.

Since all solid and liquid insulators have dielectric constants greater than air, they always tend to increase the E field in the surrounding air, making it more likely to break down. For this reason a great deal of effort is put into eliminating air (and all other gases) from insulation when fabricating bushings, capacitors, cables, and any other device which is expected to withstand high electric fields.

2.2   ELECTRIC FIELDS BETWEEN COAXIAL CYLINDERS
(Voltage Rating of Coaxial Cables)

Summary

The electrostatic fields around a cylinder can be found with very little work, and yet they are extremely important in many practical applications. In high voltage cables, for example, the relative size of the inner conductor is determined primarily by the voltage rating, since there is a critical size which gives the highest rating, regardless of the insulator, as demonstrated in this section.

Theory

We seek the field between two concentric cylindrical conductors, as shown in Figure 2.2.1. The inner conductor, which is at a potential V, has a radius a and is

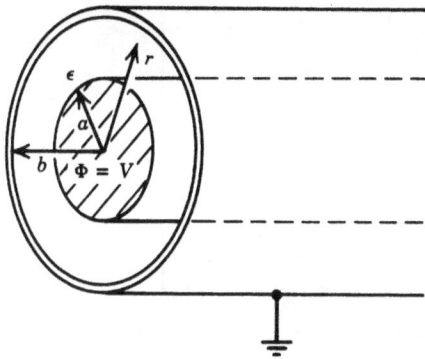

FIGURE 2.2.1. A concentric cylindrical cable.

surrounded by an insulator, which has a permittivity ϵ. The outer conductor, which is grounded, has a radius b.

The electric potential in the insulator can be obtained from the general solution of Laplace's equation in cylindrical geometry (see Appendix A). In this case there is no variation along the wire, or in the azimuthal direction, so the general solution takes the form

$$\Phi = K_1 \ln r + K_2 \qquad (2.2.1)$$

The constants can be determined from the boundary conditions on potential at the two conductors, namely,

$$\Phi(r = a) = V \qquad (2.2.2)$$

and

$$\Phi(r = b) = 0 \qquad (2.2.3)$$

Evaluating the potential at these two radii gives the solution as

$$\Phi = V \frac{\ln(b/r)}{\ln(b/a)} \qquad (2.2.4)$$

The magnitude of electric field produced by this potential

$$\mathbf{E} = -\nabla\Phi = \frac{V}{r\,\ln(b/a)}\mathbf{i}_r \qquad (2.2.5)$$

is plotted in Figure 2.2.2 as a function of radial position.

Example: Voltage Rating of Coaxial Cables

Cylindrical geometry is used in many applications because it is often easier to fabricate, or because it is easier to analyze theoretically. We review it here in the context of a common problem in high voltage design, namely, the choice of conductor size in high voltage cables. Electrical wires and cables are usually formed of a long wire or cylinder surrounded by an insulator. In many applications it is very important that the field in the insulator be purely radial, with no components in the

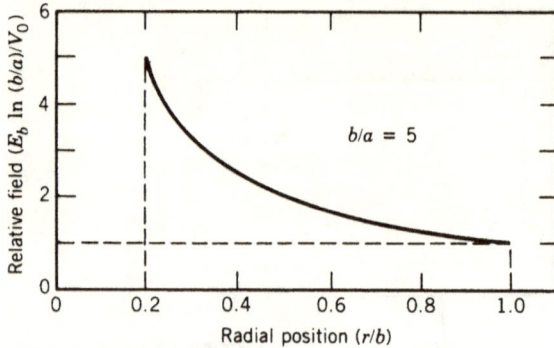

FIGURE 2.2.2.   Electric field in the coaxial cable.

azimuthal direction. For example, if the cable is insulated with the most common high voltage insulator, oil-impregnated paper, it will be extremely resistant to breakdown across the paper layers in the radial direction, but much more susceptible to breakdown along the layer. For this reason an external grounded electrode is placed outside the insulator, so that only a radial field is generated. This arrangement naturally leads to the cylindrical geometry.

Our problem is to select the size of inner conductor which gives the highest voltage rating for a given size of outer conductor.

Application of the Theory

From Figure 2.2.2 it is apparent that the maximum value of the field occurs at the surface of the inner conductor, and it is at this point that breakdown will first occur. If the breakdown strength of the insulator is E_{bk}, then the voltage rating of the cable is given from the electric field Eq. (2.2.5) as

$$V_{bk} = bE_{bk}\frac{\ln(b/a)}{(b/a)} \tag{2.2.6}$$

This voltage rating is plotted in Figure 2.2.3 as a function of the inner radius a for

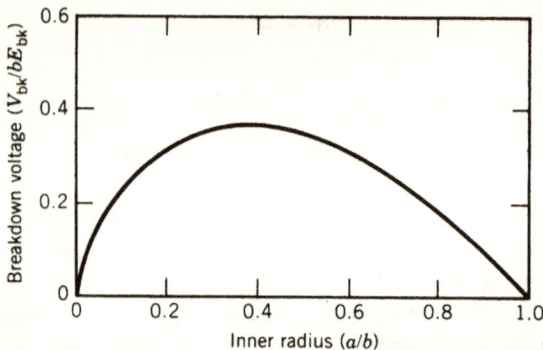

FIGURE 2.2.3.   Voltage rating of the cable as the inner radius changes.

a given outer radius and insulator strength. The curve has a maximum at a radius given by

$$\frac{\partial V_{bk}}{\partial a} = 0 \qquad (2.2.7)$$

as

$$a_{opt} = e^{-1}b \simeq 0.37b \qquad (2.2.8)$$

This choice for the inner radius gives the highest voltage rating for the cable.

Discussion

The reason for an optimum inner radius is not hard to see. If the inner conductor is very large, it will approach the outer conductor, with a very small gap between them. In this situation, a small voltage will produce a large electric field in the gap, and the cable will break down. If the inner wire is made very thin, however, the electric field near its surface will again be very large (this time because of the high curvature), and partial breakdown (or corona) will occur. At some intermediate value of inner radius, the maximum value of the electric field will be smallest, and the cable will achieve its highest rating. This value of optimum size for the inner conductor is independent of the insulator material and depends only on the size of the outer conductor. Thus all high voltage cables will be designed to approximate this size if the highest possible voltage rating must be obtained.

In practice, the cable is usually designed with an inner conductor smaller than the optimum, for reasons connected with the stability of the discharge. The thinking behind this approach can be explained by referring to the solution just obtained. If the inner conductor is smaller than the critical radius, and the voltage is raised above the breakdown level, a discharge will occur at the surface of the inner conductor. As a result, the insulator will be filled with charge carriers near the wire, and these carriers will make that region of the insulator become conducting. In effect, the inner conductor has been enlarged, since it is now composed of the wire plus the conducting region surrounding the wire. Since the inner "conductor" is now larger, the field at its outer surface is smaller, and breakdown will stop when the conducting region has expanded. This limitation of the breakdown is characteristic of corona or partial breakdown and is usually preferred in practice, since it does not lead to a complete rupture of the insulator.

If the inner conductor is at the critical radius or beyond, however, the first breakdown, which enlarges the effective radius, will expand the inner "conductor" into a region in which breakdown is even easier than before. Thus the breakdown will continue to enlarge until eventually it reaches the outer conductor. At this point, a conducting path exists from the high voltage electrode to the ground, and a large current (or short circuit) results. Since this result is far worse in practice than a partial discharge, most cables are designed with inner conductors smaller than critical radius so that the breakdown occurs in the stable, or partial discharge, regime.

As an example, consider the design of an underground power cable rated at 345 kV (488-kV peak) to be used to carry power into downtown Chicago. To establish a safety margin between the highest possible breakdown voltage and the stable or partial breakdown mode, the cable is designed so that the ratio a/b is smaller than the critical value (say $a/b \simeq 0.25$). If the breakdown strength of the insulator is 10 MV/m, Eq. (2.2.6) gives the outer radius as 141 mm and the inner radius as 35 mm.

2.3 SPHERICAL GEOMETRY
(Maximum Charge on Ink Drop Printer)

Summary

Since the sphere is the simplest three-dimensional object, it often serves as a model for an electrostatics device. One example occurs in ink jet printers, where ink drops are charged as much as possible to ensure large deflections. The maximum charge before breakdown depends on both the voltage and the electric field and is much larger than might be expected based on macroscopic breakdown measurements.

Theory

We seek the field between two concentric spheres, as shown in Figure 2.3.1. The inner sphere, which has a radius a, is assumed to be at a potential V, while the outer sphere, of radius b, is grounded ($\Phi = 0$). Just as for the coaxial geometry in the previous section, the fields can be obtained from the general solution of the potential equation, given in Appendix A. Since the arrangement here is symmetric in the angular directions, the potential will vary only in the radial direction, with the general form

$$\Phi = \frac{K_1}{r} + K_2 \qquad (2.3.1)$$

The constants can be determined from the potentials on the inner and outer spheres,

$$\Phi(r = a) = V \qquad (2.3.2)$$

$$\Phi(r = b) = 0 \qquad (2.3.3)$$

with the solution

$$\Phi = \frac{abV}{b - a}\left(\frac{1}{r} - \frac{1}{b}\right) \qquad (2.3.4)$$

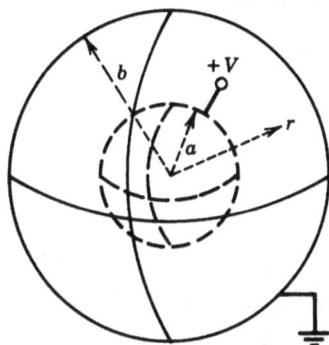

FIGURE 2.3.1. Concentric spheres.

Because the potential varies only in the radial direction, the electric field will consist of a single component,

$$E_r = -\frac{\partial \Phi}{\partial r} = \frac{ab}{b-a}\frac{V}{r^2} \tag{2.3.5}$$

As in the case of coaxial cylinders, the field also falls off away from the origin.

So far, the solution has been given in terms of the voltage on the inner sphere. In many cases, however, it is the charge on the sphere, not the voltage, which is known. This charge can be related to the voltage by Gauss' law, which takes the form

$$q = \int_{\psi=0}^{2\pi} \int_{\theta=0}^{\pi} D_r(r=a)a^2 \sin\theta\, d\theta\, d\psi \tag{2.3.6}$$

for the spherical surface at the voltage V. By using the simple permittivity relation $D = \epsilon E$, the charge can be evaluated as

$$q = CV \tag{2.3.7}$$

where C is the capacitance between the two spheres,

$$C = 4\pi\epsilon ab(b-a)^{-1} \tag{2.3.8}$$

Example: Maximum Drop Charge in an Ink Jet Printer

In an ink jet printer, small drops of ink receive a charge and are then deflected by an electric field to an appropriate point on the paper. To obtain maximum deflection, the charge should be as high as possible, but it can not be so high that corona discharge occurs, since this would decrease the original charge, and change the drop trajectory from its desired value.

Before discharge can occur, two conditions must usually be met. Both the voltage drop and the electric field must exceed critical values. The maximum values for each of these quantities depend on the charge on the drop, which can be calculated from the theoretical results just obtained.

Application of the Theory

The breakdown conditions in air can often be written as

$$E > E_{bk} \simeq 3 \text{ MV/m} \tag{2.3.9}$$

and

$$V > V_{bk} \simeq 300 \text{ V} \tag{2.3.10}$$

The approximate values given are based on measurements on parallel plate arrangements and are not necessarily valid for a spherical drop. They are adequate for an estimate, however, and are used here.

To calculate the field and potential at the drop surface, we need to know the location of the other electrode. In fact, the model also assumes that this electrode is spherical, although the geometry inside an ink jet printer is always much more complex than this. In practice, the location of the outer electrode is not very important, since it is usually far away compared to the size of the drop. In the spherical geometry, unlike the cylindrical, the fields approach a well-behaved limit as the outer electrode is removed, so its exact location (and shape) is immaterial if it is far enough away ($b \gg a$). Under these conditions the voltage on a drop with charge q is from Eq. (2.3.7),

$$V = \frac{q}{4\pi\epsilon a} \tag{2.3.11}$$

The maximum electric field around the drop occurs at the surface ($r = a$) and is given by Eq. (2.3.5) as

$$E_m = \frac{V}{a} = \frac{q}{4\pi\epsilon a^2} \tag{2.3.12}$$

The limits on voltage and field can now be translated into charge limits by giving the two necessary conditions for breakdown as

$$q > 4\pi\epsilon a V_{bk} \tag{2.3.13}$$

and

$$q > 4\pi\epsilon a^2 E_{bk} \tag{2.3.14}$$

The charge required to satisfy these conditions is shown as a function of drop size in Figure 2.3.2. The charge is limited by electric field for large drop sizes while it is limited by voltage for small drops. The demarcation between these two regimes occurs at the critical size determined by the intersection of the two charge equations, given by

$$a_{cr} = \frac{V_{bk}}{E_{bk}} \simeq 100 \ \mu\text{m} \tag{2.3.15}$$

In an ink jet printer the drops are usually much smaller than this, so it is the voltage, not the maximum field, which sets the limit on the charge. For a typical

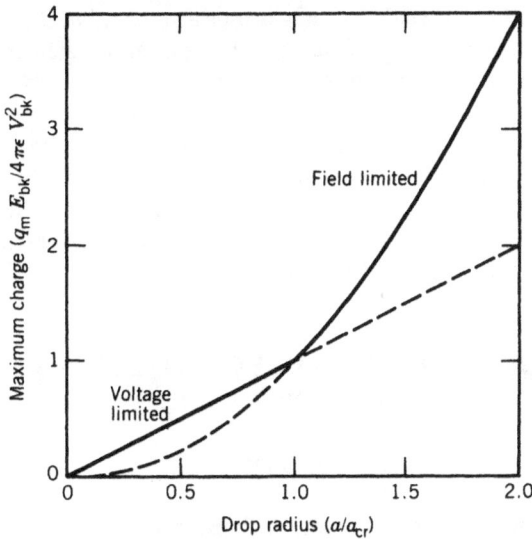

FIGURE 2.3.2. Limiting charges for voltage and field breakdown from a charged sphere.

drop with a radius of 20 μm, the maximum charge would be $q < 0.67$ pC. At this value the field at the surface would be $E = 15$ MV/m.

Discussion

The high value of electric field in this application, far above the value which causes breakdown in macroscopic applications, is typical of very small devices. It is the main reason why electrostatics tends to become more important on small size scales, where the fields, and hence the forces, can be relatively high. In the ink jet printer the high values of charge allow larger deflections of the ink drops, so that a single stream of drops can cover a large area of paper.

Although electrical breakdown must always be considered in an application like this, there is an additional effect which can occur when a liquid drop receives a high charge. The electrostatic repulsion of the charge, which resides on the surface of a conducting drop, acts to pull the drop apart, usually by stretching it out as the charges flow to opposite ends (Melcher, 1981). The surface tension of the drop opposes the electrostatic repulsion, but if the charge on the drop is large enough, it will eventually overcome surface tension, and the drop will split. For water, the charge at which this breakup occurs is roughly equivalent to the charge for breakdown, so that both processes must be considered in the design of a printer.

When the grounded electrode is far from the charged sphere, the relation between charge and voltage approaches the limiting form given in Eq. (2.3.11), implying a capacitance of

$$C = 4\pi\epsilon a$$

This value is often quoted as the capacitance of a sphere, but it should be kept in mind that there is no such thing as the capacitance of an isolated conductor. It just happens in this case that the value of the capacitance between the inner and outer electrodes is very insensitive to the actual position of the outer electrode when it is far away. If the sphere is picked up and brought closer to another conductor, the capacitance, electric field, and potential will all change, and this will in turn affect the maximum charge allowed before breakdown occurs. Since an ink drop ends its flight by contacting a grounded surface (the paper), conclusions drawn from the approach just given can only be first approximations.

2.4   CYLINDERS IN EXTERNAL FIELDS
(St. Elmo's Fire)

Summary

An object placed in an electrostatic field will distort the field and acquire a charge distribution. Both of these effects often have practical significance. St. Elmo's fire, for example, is a corona discharge which occurs when cylindrical lines and spars of a ship (or airplane) enter the high atmospheric electrical fields near storms.

Theory

The simplest model for a material placed in an external field is the circular cylinder shown in Figure 2.4.1. Far away from the cylinder the electric field is uniform and constant, with a value of E_0. In the figure the field is directed along the horizontal z axis to simplify the equations, but any orientation can be chosen to match the actual situation. Near the surface of the cylinder the field is distorted, strengthening it in some areas and weakening it in others.

The distortion in the field can be found from the general solution of Laplace's equation (Appendix A), but there are an infinite number of such solutions, and some thought must be given to selecting the appropriate ones for each case. Here, the appropriate geometry must be selected first, since the model contains both a

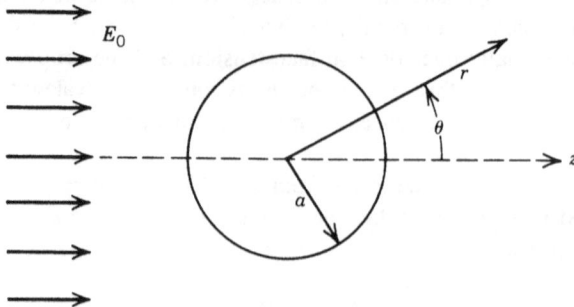

FIGURE 2.4.1.   A cylinder in external field.

rectangular component (the external field) and a circular component (the cylinder). In general it is wise to pick the geometry to fit the model in its most complex region, which would be the curved surface of the cylinder. Thus a circular coordinate system is depicted in the figure.

Even in this simple geometry, there are still many possible solutions. Usually, the choice is guided by the boundary conditions. The only source of electric field here is the external field, which can be written as

$$\mathbf{E} = E_0 \mathbf{i}_z \qquad (2.4.1)$$

corresponding to the external potential

$$\Phi = -E_0 z \qquad (2.4.2)$$

in rectangular coordinates. Since we have selected circular coordinates for this model, however, the external potential should be rewritten as

$$\Phi = -E_0 r \cos \theta \qquad (2.4.3)$$

far from the surface of the cylinder ($r \rightarrow \infty$). This boundary condition requires that we select from the general solution [Eq. (A.12)] only terms which vary with $\cos \theta$ so as to match the external field far from the cylinder,

$$\Phi = K_1 r \cos \theta + K_2 r^{-1} \cos \theta \qquad (2.4.4)$$

At the surface of the conducting cylinder the tangential electric field will vanish,

$$E_\theta = -\frac{1}{r} \frac{\partial \Phi}{\partial \theta} = \left(K_1 + \frac{K_2}{a^2} \right) \sin \theta = 0 \qquad (2.4.5)$$

which sets one restriction on the two constants K_1 and K_2. A second restriction comes from matching the field far from the cylinder,

$$\Phi(r \rightarrow \infty) = K_1 r \cos \theta = -E_0 r \cos \theta \qquad (2.4.6)$$

Combining the two conditions gives the potential everywhere outside the cylinder as

$$\Phi = -E_0 \left(r - \frac{a^2}{r} \right) \cos \theta \qquad (2.4.7)$$

with the electric field components

$$E_r = -\frac{\partial \Phi}{\partial r} = E_0 \left(1 + \frac{a^2}{r^2} \right) \cos \theta \qquad (2.4.8)$$

$$E_\theta = -\frac{1}{r} \frac{\partial \Phi}{\partial \theta} = -E_0 \left(1 - \frac{a^2}{r^2} \right) \sin \theta \qquad (2.4.9)$$

Example: St. Elmo's Fire

A traditional component of seafaring tales is St. Elmo's fire (Newman and Robb, 1977), the mysterious light which often appeared along the rigging of sailing ships near storms. This is a manifestation of field distortion effect just described, with the ropes and spars playing the role of the circular conductor immersed in an external field. The field in this case is the atmospheric field, which is normally quite weak, but can attain very high values in the vicinity of thunderstorms. Even though the field is not strong enough to cause the massive breakdown associated with lightning, it may be intensified near the cylindrical surfaces to the point where a local discharge, or corona, may occur. St. Elmo's fire also occurs on airplanes, where it can disrupt radio communications.

Application of the Theory

In this application the radial electric field at the surface of the cylinder is most important, since this is where the electric field reaches its greatest magnitude. It varies around the surface and is given by

$$E_r = 2E_0 \cos \theta \tag{2.4.10}$$

with its extreme values along the axis. From the equation, it is clear that the field at the surface of the cylinder may be double the field in the distance, so that breakdown can occur near a spar when it might be impossible in free space.

The field is intensified near the curved surface, and it might be expected that the intensification is greater for the smaller cylinders, which are more highly curved. Surprisingly, this is not true. A glance at the equation for the field at the surface of the cylinder, Eq. (2.4.10), shows that the field is doubled regardless of the size, so that all spars and lines exposed to the external field are likely to go into corona together. In fact, this behavior is the rule, with variations more likely to be caused by the variation of the gross electric field in the vicinity of the ship rather than by the distribution of sizes of the individual elements.

Discussion

An electric field which does not increase with curvature is a rarity in electrostatics, and is worthwhile to examine how this type of behavior arises. In most cases, a body with a given charge or voltage has a field at the surface which increases as the body becomes smaller (more curved). In this example, however, the cylinder has no net charge, and the field distortion arises from the charge induced by the external field. In effect, the cylinder is shorting out part of the external field, roughly equivalent to the voltage drop across the diameter of the cylinder. A larger cylinder intercepts more field, and therefore induces a greater electric dipole charge. The increase in induced charge for the larger cylinder offsets the decrease in field caused by the decreased curvature of the larger cylinder, and the field distortion remains constant for any size.

In the theoretical model presented here, as in many such models, there are a number of unstated assumptions which may affect the application of the model to a given situation. One of these assumptions concerns the selection of solutions to the potential equation. We did not include the term K/r used in the previous section because we did not need it to match the stated boundary conditions. Left unstated, however, was an additional condition, namely, the net charge and the potential of the cylinder were both zero. While this might be acceptable in some manifestations of St. Elmo's fire, such as in airplanes and dirigibles, it is open to question for ships in a storm, which are in contact with the grounded sea and might carry a net induced charge, much like a lightning rod. In such cases, the extra term should be added to the solution.

2.5   SPHERES IN EXTERNAL FIELDS
(Bubble Breakdown in Insulating Oils)

Summary

A sphere in an external field is the model used most often for three-dimensional particles in external fields. The basic solutions are similar to the two-dimensional cylinder just considered, but the inclusion of more complex boundary conditions allows these solutions to represent a wide variety of additional application, such as bubble breakdown in insulating oil.

Theory

A sphere of radius a is immersed in an external field, as shown in Figure 2.5.1. Both the sphere and the external medium are conducting, with ohmic conductivities σ_a and σ_b. As in the previous section, the external field is oriented along the horizontal axis, which is also the polar axis of the spherical coordinate system. The fields outside the sphere are similar to those described for the cylinder. In particular, the potential outside the sphere has the form

$$\Phi_b = -E_0 r \cos \theta + K_1 r^{-2} \cos \theta \tag{2.5.1}$$

which is taken from the more general solution of Eq. (A.19).

In this case the bubble is not a conductor, so there may be an electric field inside the sphere. The process of selecting suitable terms for this region must begin all over again, based on the boundary conditions appropriate to the inside of the

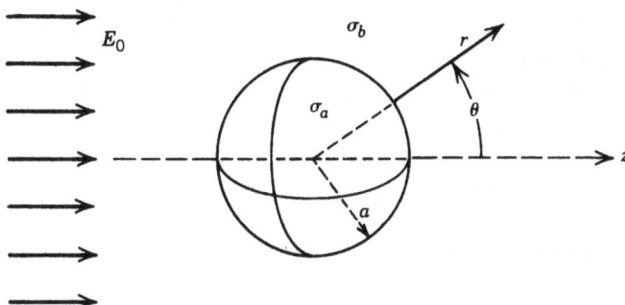

FIGURE 2.5.1.   A sphere in an external field.

sphere. Assuming that there is no initial charge inside the sphere, the potential must be finite, and all terms with inverse powers of r can be discarded. Also, since the potentials inside and outside must both satisfy the boundary conditions at the surface of the sphere, they will need to have the same θ dependence, so the potential inside the sphere must have the form

$$\Phi_a = K_2 r \cos \theta \qquad (2.5.2)$$

The two constants K_1 and K_2 are determined by boundary conditions at the surface of the sphere, which can become quite subtle in practice owing to the presence of double layers and charge relaxation (Melcher, 1981). In a relatively good insulator, however, double-layer effects are usually small, so the boundary condition for the tangential electric field becomes

$$E_\theta(r = a^+) = E_\theta(r = a^-) \qquad (2.5.3)$$

Charge relaxation at the interface must always be considered, however, since it is a dynamic process and occurs for at least part of the time after any change in the electric variables. It is described by the second boundary condition in terms of the current which flows to the surface and the surface charge which builds up there as

$$\sigma_a E_r(r = a^-) - \sigma_b E_r(r = a^+) = \frac{d}{dt}[\epsilon_b E_r(r = a^+) - \epsilon_a E_r(r = a^-)]$$

$$(2.5.4)$$

for an ohmic conductor, where Gauss' law has been used to express the surface charge in terms of the electric fields. This boundary condition is not as simple as the previous ones since it depends on the exact time behavior of the applied fields. The effect of changes in this behavior is studied in more detail later in this book; for now we concentrate on two limiting forms in which either ohmic conduction or surface charging can be completely neglected. These two cases are usually called the dc and ac (or high frequency) limits, respectively.

In the dc limit, variations, if any, in the surface charge are so slow that no net ohmic current through the liquid is needed to charge the interface. In terms of the boundary conditions, this amounts to neglecting the time derivative terms, so that

$$\sigma_a E_r(r = a^-) = \sigma_b E_r(r = a^+) \qquad (2.5.5)$$

In this limit the radial (or normal) currents must be continuous across the interface. Since the conductivity is different in the two regions, the electric fields must also differ. Using the two boundary conditions of Eqs. (2.5.3) and (2.5.5) with the external and internal solutions of Eqs. (2.5.1)–(2.5.2) gives the potential in the sphere as

$$\phi_a = -E_0\left(\frac{3\sigma_b}{\sigma_a + 2\sigma_b}\right) r \cos \theta \qquad (2.5.6)$$

and in the external medium as

$$\phi_b = -E_0\left[r - \frac{a^3}{r^2}\left(\frac{\sigma_a - \sigma_b}{\sigma_a + 2\sigma_b}\right)\right] \cos \theta \qquad (2.5.7)$$

In the high frequency limit, the external field is varying too quickly to allow the conduction process to change the charge at the surface of the sphere, so that

$$\frac{d}{dt}[\epsilon_b E_r(r = a^+) - \epsilon_a E_r(r = a^-)] = 0 \tag{2.5.8}$$

This condition requires only that the charge at the interface remain constant. With a simple interface, however, it is also true that the net charge vanishes, so that the constant value may be set equal to zero. If this is not true, we can restrict our attention to the field components induced by the external field, leaving the net charge contribution to a separate calculation. In either case, the boundary condition becomes

$$\epsilon_a E_r(r = a^-) = \epsilon_b E_r(r = a^+) \tag{2.5.9}$$

and the potential inside the sphere is

$$\Phi_a = -E_0 \left(\frac{3\epsilon_b}{\epsilon_a + 2\epsilon_b}\right) r \cos \theta \tag{2.5.10}$$

while the potential in the external medium is

$$\Phi_b = -E_0 \left[r - \frac{a^3}{r^2} \left(\frac{\epsilon_a - \epsilon_b}{\epsilon_a + 2\epsilon_b}\right) \right] \cos \theta \tag{2.5.11}$$

Note that the form is quite similar to the dc case, with the permittivities of the two regions replacing the conductivities. This does not mean that the fields are similar in the dc and ac limits, however, since the relative permittivities and conductivities usually differ considerably for any two materials.

Example: Bubble Breakdown in Insulating Oil

Insulating oil used in transformers and cables is often contaminated by small gas bubbles which may develop during the life of the device, especially if excessive voltages are applied. These bubbles distort the electric field, which makes electrical breakdown more likely. The theory just developed can be used to find the largest increase in the electric field and the location at which it occurs. Since bubbles can be expected in any device, this allows us to estimate the safety margin needed to prevent bubble breakdown in the device.

Application of the Theory

At 60 Hz the conduction inside an insulating oil is, by definition, small, so the ac limit is appropriate for this problem. The presence of the bubble will distort the field in both the oil and the bubble. Whenever the field is distorted, there will be regions in which the magnitude of the field is increased, and this could occur both in the oil and the gas bubble. The effect is much more important in the gas however, since the breakdown fields in gases are usually much lower than in oils.

In the bubble (the internal medium) the potential is given from Eq. (2.5.10) as

$$\Phi_a = \frac{-3\epsilon_b}{\epsilon_a + 2\epsilon_b} E_0 z \qquad (2.5.12)$$

by converting back to rectangular coordinates. The electric field is much easier to visualize in the rectangular coordinate system, where it assumes the same simple form as the external field. The electric field is uniform inside the bubble, with a magnitude which depends only on the relative permittivities of the oil and bubble. The dielectric constant of gases is usually quite close to unity, whereas most insulating oils have dielectric constants near 2.5. Using these numbers, we find the field inside the bubble is stronger than the externally applied field by a factor of 25%. For example, a field in the oil of 5 MV/m would cause a field inside the bubble of 6.25 MV/m. As in the example of St. Elmo's fire, the increase in the field is independent of the size of the bubble, so small bubbles will be stressed as much as larger ones.

A field increase of this magnitude would not be too serious if it occurred in the oil, since a transformer would not usually be designed to operate within 25% of the breakdown field. Unfortunately, this field occurs in a gas, which will almost always have a much lower breakdown field than the oil. (Typical values are near 3 MV/m.) Thus bubble formation almost invariably leads to breakdown within the insulating oil.

Discussion

Bubble breakdown is not catastrophic, since it is localized in the bubble, and does not bridge the insulating oil gap. Over the long term, however, the chemical products of the breakdown lead to deterioration of the insulating properties of the oil, so partial breakdown of this sort is best avoided in practice. When the device is constructed, the oil can be degassed by heating in a vacuum, and the container can also be evacuated to remove any gases which might become dissolved in the liquid. Gases may still evolve during operation, however, especially in the high fields which develop near small impurity particles. These fields, by the way, can also be computed using the theory of this section.

BIBLIOGRAPHY

Chalmers, J. A., *Atmospheric Electricity*, 2nd ed., Pergamon, Oxford, 1967.

Cobine, J. D., *Gaseous Conductors*, Dover, New York, 1958.

Graneau, P., *Underground Power Transmission*, Wiley, New York, 1979.

Ink Jet Printers, *IBM J. Res. Devel.*, **21** (1) 1–96 (Jan. 1977).

Melcher, J. R., *Continuum Electromechanics*, MIT Press, Cambridge, MA, 1981.

Newman, M. M., and J. D. Robb, Protection of aircraft, in *Lightning*, R. H. Golde, Ed., Academic, New York, 1977, Vol. 2, Ch. 21.

Peek, F., *Dielectric Phenomena and High Voltage Engineering*, McGraw-Hill, New York, 1920.

PROBLEMS

PROBLEM 1 (ELECTRIC POWER)

a.   A wire carrying high voltage V_0 is brought through a grounded circular air-filled hole into a power substation. What is the maximum voltage which can be applied to the wire before the breakdown strength of the air (3×10^6 V/m) is exceeded? The wire has a radius $a = 1$ mm and is centered in the hole, which has a radius $b = 3$ cm.

b.   In an effort to increase the voltage the wire can carry, a porcelain insulator of radius $b = 1$ cm is fitted snugly over the wire. The porcelain has a dielectric constant $\epsilon/\epsilon_0 = 7$ and a breakdown strength $E_m = 6$ MV/m. What is the maximum voltage which can now be applied to the wire without causing breakdown anywhere within the channel? This is an example of insulation coordination, which is an important part of insulator design.

PROBLEM 2 (ATMOSPHERIC ELECTRICITY)

During fair weather there is a steady vertical electric field in the atmosphere, E_0 ($= 100$ V/m). If a fog forms, the field is altered because fog has a lower conductivity than clear air. If the external field E_0 remains constant, find the field $E(t)$ in a horizontal fog layer of depth a ($= 100$ m) which forms very quickly at $t = 0$. The conductivity of the clear air is σ_a ($= 20 \times 10^{-15}$ S/m) whereas that of fog is σ_b ($= 10 \times 10^{-15}$ S/m).

PROBLEM 3 (ELECTRIC POWER)

Underground power cables consist of a conductor of radius a ($= 1$ cm) covered by an insulator of radius c ($= 3$ cm), which is in turn covered by a grounded conductor. In operation, the inner conductor is heated by joule losses, which makes the inner portion $a < b$ ($= 2$ cm) of the insulator hotter than the outer portion. This changes the conductivity and permittivity throughout the cable insulation. This distribution is modeled by two concentric regions. In the inner region ($a < r < b$),

$$\sigma = \sigma_a \ (= 1 \text{ nS/m}), \qquad \epsilon = \epsilon_a \ (= 20 \text{ pF/m})$$

while in the outer region ($b < r < c$)

$$\sigma = \sigma_b \ (= 0.5 \text{ nS/m}), \qquad \epsilon = \epsilon_b \ (= 20 \text{ pF/m})$$

a.   If the inner conductor is held at a dc voltage V ($= 100$ kV), find the electric fields in the two layers of the insulator.

b.   Is the maximum field greater or less than that in a uniform (unheated) insulator?

PROBLEM 4   (INSTRUMENTS)

Van der Graaf generators are usually surmounted by a spherical cap to reduce the electric field strength and lessen corona losses. What diameter cap is needed for a 10-MV generator, assuming that the ground potential is infinitely far away? How much charge is stored in the cap? (Use $E_m = 3$ MV/m as the corona onset field.)

PROBLEM 5   (ELECTRIC POWER)

Insulating oil used in transformers and cables is occasionally contaminated with small metallic particles due to machining operations or careless assembly. These particles distort the field in the oil, leading to partial breakdown (corona) at their surface.

  a.   If one of these particles is modeled by a sphere of radius a in an initially uniform electric field E_0, find an expression for the maximum value of electric field in the oil.

  b.   Commercial transformer oil having the properties $\sigma = 10$ pS/m, $\epsilon = 2.5\ \epsilon_0$, $E_m = 10$ MV/m is used in a transformer in which the applied field $E_0 = 5$ MV/m. Will breakdown occur?

PROBLEM 6   (COMPUTER PERIPHERALS)

In an ink jet printer the drops are charged by surrounding the jet of radius a ($= 15\ \mu m$) with a concentric cylindrical electrode of radius b ($= 1$ mm) which is held at a voltage V. What voltage is needed to generate a charge of q ($= 20$ fC) on the drop formed from a section of the jet of length λ ($= 100\ \mu m$)?

3

FIELDS CAUSED
BY CHARGES

The presence of charges complicates the electric field solution in three ways. Since the solutions of Laplace's equation are useful only when lines of constant voltage correspond to the physical boundaries, the presence of charge on the boundaries may make the application of those solutions impossible. In any case, we are required to examine the effect of the charges on the fields at the boundaries, usually with the aid of Gauss' law. This is exemplified by a problem which arises in xerographic copiers.

A second complication arises if the charges exist within the space under consideration, and not just at the boundaries. Space charge requires the full Poisson equation for the fields, with solutions which are more involved than those obtained with the homogeneous Laplace equation. One of these fields, which arises in connection with grain elevator explosions, is considered in Section 3.2.

Finally, the charges are often free to move, and this motion induces charge rearrangement throughout the entire region. This accounts for the current which flows in the external circuitry of many electrostatic devices, such as the photo-tube.

3.1 FIELDS CAUSED BY SURFACE CHARGES
(Xerographic Development)

Summary

Electrostatic problems in which the charge is specified are very common in many applications, as well as in nature. We begin the study of this type of problem with an example in which a surface is charged by external means and the resulting electric fields must be found. The context here is the development step in a xerographic copier, in which the electric fields generated by photo-controlled discharges attract the pigment particles which will eventually form an image on the paper.

Theory

Until now the effect of charges on the interface between two materials was neglected, although it was mentioned that this charge is often very important. This section includes the effect of surface charge distributions in the analysis. Consider a surface charge layer, between two insulators, as shown in Figure 3.1.1. The potential in each of the two insulating layers is governed by Laplace's equation because there are no volume charges. Just as in Section 2.1, the potentials in the insulators above ($_a$) and below ($_b$) the surface are given by

$$\Phi_a = K_{a1}x + K_{a2} \tag{3.1.1}$$

$$\Phi_b = K_{b1}x + K_{b2} \tag{3.1.2}$$

At the two grounded electrodes, boundary conditions are given by

$$\Phi_a(x = a) = 0 \tag{3.1.3}$$

and

$$\Phi_b(x = -b) = 0 \tag{3.1.4}$$

At the surface of the photoconductor the potential must be continuous, as before,

$$\Phi_a(x = 0) = \Phi_b(x = 0) \tag{3.1.5}$$

while the difference in D fields is related to the charge on the surface by

$$D_a - D_b = \rho_s \tag{3.1.6}$$

When the field solutions are combined with the definition of the D field

$$D = \epsilon E \tag{3.1.7}$$

the electric fields above and below the charge layer are given by

$$E_a = \frac{\rho_s}{\epsilon_a}\left(1 + \frac{\epsilon_b}{\epsilon_a}\frac{a}{b}\right)^{-1} \tag{3.1.8}$$

$$E_b = -\frac{\rho_s}{\epsilon_a}\left(\frac{a}{b}\right)\left(1 + \frac{\epsilon_b}{\epsilon_a}\frac{a}{b}\right)^{-1} \tag{3.1.9}$$

Example: The Xerographic Development Electrode

One of the most important applications of electrostatics is the xerographic copying machine (Schaffert, 1966), which forms an electrostatic image in the following

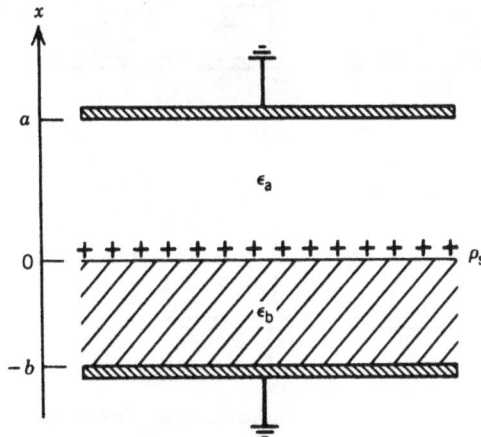

FIGURE 3.1.1.   A charged surface between two electrodes.

manner. Initially the entire surface of a photoconductor is charged uniformly, as shown in Figure 3.1.2. The charge on the top surface comes from ions produced by a corona discharge in the air above the photoconductor, while the opposite charge on the lower surface comes from the metal below, drawn by attraction to the upper charges. Since the photoconductor is an insulator in the dark, the lower charges can get no farther than the bottom of the photoconductor.

When light from the document to be copied is focused on the photoconductor, it becomes a conductor, and the charges on the lower surface can combine with those on the upper surface to neutralize each other. The image is then developed by pouring a charged black powder over the surface of the photoconductor. The powder is attracted by the electric field of the charged surface. Later, the powder will be transferred to paper and melted to form a permanent image.

Application of the Theory

Since the image is formed by the attractive forces of the electric field, the quality of the image depends crucially on the strength of this field, which is easy to calculate if the surface charge is uniform, as it might be when the image contains wide areas of uniform darkness. This situation can be described by the foregoing theory.

The magnitude of the surface charge is determined by the nature of the charge source, the geometry of the machine, and other parameters discussed in later chapters. For now, we assume its value is given. The electric field in the air above the charged photoreceptor is given by Eq. (3.1.8), which can be rewritten as

$$E_a = \frac{\rho_s}{\epsilon_a} \left[1 + \frac{a}{(b/\kappa)} \right]^{-1} \qquad (3.1.10)$$

where κ is the dielectric constant of the photoconductor. This field is plotted in Figure 3.1.3 as a function of the distance between the charged surface and the upper electrode.

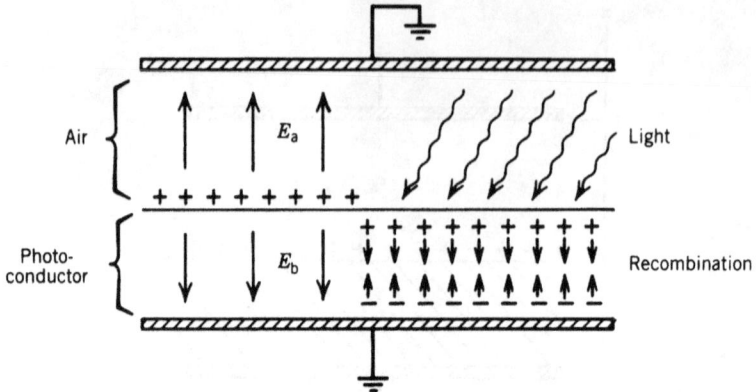

FIGURE 3.1.2. Light discharges a part of the surface.

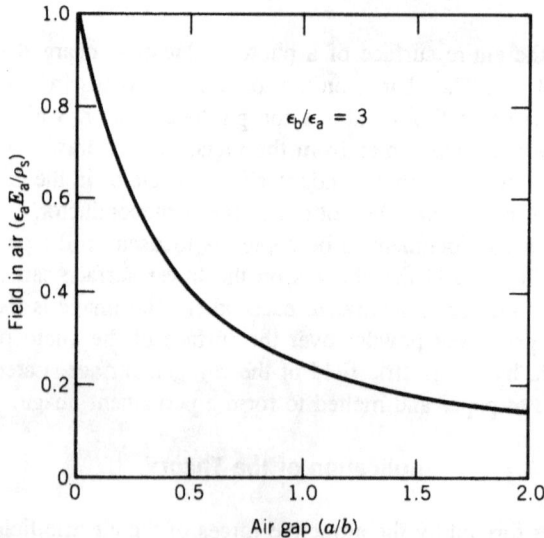

FIGURE 3.1.3. Electric field in the air near a development electrode.

Discussion

From the figure it is clear that the field is very weak in the air when the ground plane is far from the surface. This is the most common situation and is usually assumed if no other indication is given. Because the external field is weak, the pigment powder is not attracted very strongly, and it will not stick to the photoreceptor. As a result, the copy will show only a faint image where the original was uniformly dark. Failure of solid area coverage by this mechanism was very common in early xerographic copiers.

Near the edge of a solid area, or at the image of a thin line this effect does not occur because fringing fields are strong there, and the assumption of uniform fields is not valid. Since most copying is done on typed or printed originals in which the image consists exclusively of thin lines, failure of solid area coverage is not too serious and is sometimes ignored in the design. If solid area coverage is needed, the results illustrated in Figure 3.1.3 show how it can be obtained. The electric field in the air becomes larger as the grounded electrode is moved closer to the surface of the photoconductor. Inclusion of an electrode close to the surface (called a development electrode) will increase the external field and lead to a greater collection of pigment powder on the surface of the photoreceptor and thus to a darker and more uniform image. The extension of this technique, in the form of a magnetic brush, is used in most copiers which require a high quality image.

From this example it is clear that the location of the upper electrode plays a key role in the field distribution, but in many discussions it is not even mentioned. This omission often leads to confusion over the assumptions implied by its absence. In reality, there is always *something* which serves as the second electrode, even if it is only the lid of a machine or the wall of a room.

The relative effect of the electrodes is often described in terms of a *dielectric thickness*, which is based on Eq. (3.1.10). There, the thickness of the photoreceptor appears only as the ratio b/κ, which gives the effective thickness of the dielectric based on its influence on the field distribution. This dielectric thickness is always less than the actual thickness, which implies that the electrode on the other side of the dielectric has more influence on the fields than the geometry would suggest.

3.2 ELECTRIC FIELDS WITH SPACE CHARGE
(Grain Elevator Explosions)

Summary

When the charge which produces the field is not confined to surfaces, the field problem is somewhat more difficult, since the space-charge term appears in the general solution and not just in the boundary conditions. Such fields are extremely important in many areas, however, especially when transporting large amounts of combustible insulators, such as grain and oil. Many of the explosions which occur in these industries can be traced to sparks produced by the accumulation of charge and the accompanying high electric fields.

Theory

The electric fields and voltages caused by volume charges can be determined fairly easily if the charge distribution is known in advance. Often the charge density will be approximately uniform so that

$$\rho = \rho_0 \qquad (3.2.1)$$

and Poisson's equation takes the form

$$\nabla^2 \Phi = -\frac{\rho_0}{\epsilon} \tag{3.2.2}$$

In many applications the charge is distributed throughout a circular cylinder pipe as depicted in Figure 3.2.1. Since the charge density does not vary along the length of the pipe, neither will the potential, and this equation simplifies to

$$\frac{1}{r}\frac{d}{dr}\left(r\frac{d\Phi}{dr}\right) = -\frac{\rho_0}{\epsilon} \tag{3.2.3}$$

The presence of the space-charge term makes the solution of Poisson's equation more difficult than the solution of the corresponding Laplace's equation. The most useful approach to such an inhomogeneous differential equation involves superposition of a particular solution related to the space charge with as many solutions of the homogeneous (Laplace) equation as are needed to match the boundary conditions. For the uniform charge distribution of Eq. (3.2.3), successive integration with respect to r gives

$$r\frac{d\Phi}{dr} = \frac{-\rho_0 r^2}{2\epsilon} + K_1 \tag{3.2.4}$$

and then

$$\Phi = \frac{-\rho_0 r^2}{4\epsilon} + K_1 \ln r + K_2 \tag{3.2.5}$$

Note that the second and third terms are the solutions of Laplace's equation appropriate to the cylindrical symmetry. The constants K_1 and K_2 are determined by the conditions that the pipe is grounded, $\Phi(a) = 0$, and that the potential is finite everywhere inside the pipe, in particular at $r = 0$. This gives the potential and field inside the pipe due to the space charge as

$$\Phi = \frac{\rho_0 a^2}{4\epsilon}\left(1 - \frac{r^2}{a^2}\right) \tag{3.2.6}$$

$$\mathbf{E} = -\nabla\Phi = \frac{\rho_0 r}{2\epsilon}\mathbf{i}_r \tag{3.2.7}$$

These results are shown in Figure 3.2.2.

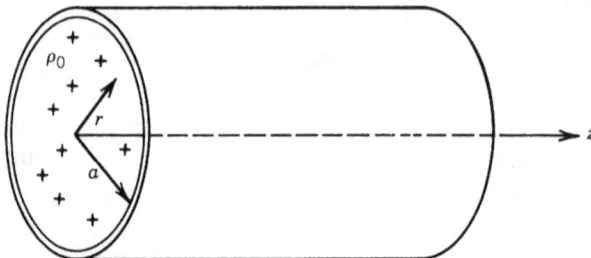

FIGURE 3.2.1. Circular pipe filled with charge.

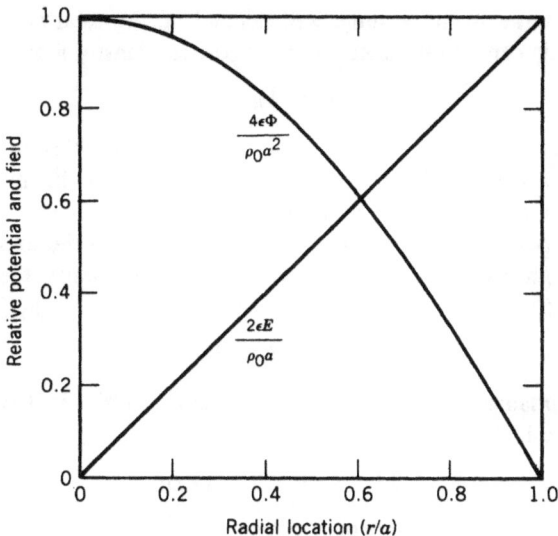

FIGURE 3.2.2. Electric field and potential inside the pipe.

Example: Grain Explosions

One important instance of charges distributed in space is the transport of insulating materials such as oil and grain through pipes. The relative movement of the material past the wall of the pipe leads to charging, just as walking across a rug in winter does, and these charges are then carried along the pipe by the mechanical force of the flow. Several explosions and fires in grain processing and oil tankers have been attributed to sparking caused by this charging effect. Whether such discharges occur depends on the strength of the electric fields created by the space charge.

Application of Theory

The space-charge field solutions are useless in practice unless the charge density is known inside the pipe, and this is the quantity that is often difficult to determine in advance when designing safe systems for the transport of insulators such as grain. A rough estimate for the charge density can be obtained, however, based on the particle density and the maximum surface charge which can be sustained before breakdown. If the grain particle picks up charge of a preferred sign from the wall, it can only increase its charge as it flows along, especially if the conductivity of the surrounding medium is too low to permit charge to leak off. Eventually, however, the charge increases to the point that the electric field between the particle and the wall allows local discharges back to the pipe after contact. The limiting surface charge in air is 25 $\mu C/m^2$. A reasonable estimate for the average charge density on the powder surface would be somewhat smaller than this, so we assume that $\rho_s \simeq 10\ \mu C/m^2$.

The total charge in the volume will depend on the total surface area in the volume, and this can be estimated from the particle density n and radius b as

$$A \simeq 4\pi b^2 n \qquad (3.2.8)$$

For a typical grain elevator, $n \simeq 10^{12}$ m^{-3} (i.e., the particles are approximately 100 μm apart) and $b \simeq 10$ μm, so the total area is 1.26×10^3 m^2, and the space charge density can be estimated as 12.5 mC/m^3.

The largest potential in the pipe will be reached along the axis, which is farthest from the grounded wall. For a pipe with a 10-cm radius, the value is given from Eq. (3.2.6) as

$$\Phi(0) = 3.55 \text{ MV} \qquad (3.2.9)$$

which is a surprisingly high value. The maximum electric field, which occurs at the wall, is also high,

$$E(r = a) = 71 \text{ MV/m} \qquad (3.2.10)$$

far greater than that required for breakdown.

Discussion

Under the conditions assumed here, it is obvious that electric discharges can be expected quite often when pumping insulating powders such as grain through pipes. Whether such discharges lead to explosions or fires is determined by the ignition energy of the material, the fuel/air ratio, and other factors. Both the maximum voltage and the maximum field increase with the size of the container, so discharges are even more likely to occur in larger containers.

The direct integration of Poisson's equations, as demonstrated here, is most useful when the charge distribution depends on, at most, one variable and is not influenced by the electric field. In practice, neither of these restrictions is likely to hold, and the field solutions quickly become complicated. The simpler model presented here is most useful in preliminary work, where it gives insight into the nature and magnitude of the electrostatic effects.

3.3 FIELDS AND CURRENTS FROM MOVING CHARGE
(Radiation Detectors)

Summary

When charges move, the electrostatic fields are still calculated in the same way as when they are stationary, but the currents induced in the surrounding materials must be determined by careful analysis. One of the commonest occurrences in electrostatic applications is the approach of a charge to a conductor. This apparently simple situation can give rise to some unexpected results, as we demonstrate in this section in the context of a radiation detector.

Theory

In many applications of electrostatics the fields are changing in time as a result of the movement of the charges which produce them. As the charges move, they generate a current flow in the external circuit. To find the current flow, it is usually best to make use of the conservation of charge equation,

$$\frac{dq}{dt} + i = 0 \qquad (3.3.1)$$

applied at the electrode to which the desired circuit current leads. The charge on this electrode is given by Gauss' law as

$$q = \oint \mathbf{D} \cdot d\mathbf{A} \qquad (3.3.2)$$

so that the current can be determined as soon as the electric field in the adjacent insulator is known.

Example: A Phototube

One example of this class of problem is a radiation detector (e.g., a phototube) in which a burst of radiation striking a metallic electrode leads to the emission of charges (usually electrons), which then move through an insulating region to another electrode. As an example of the response of such a detector, consider the current produced by a localized flash of light. The charges resulting from photoelectric emission are attracted toward the opposite electrode, so that there is a packet of charge at the position z at some later time, as shown in Figure 3.3.1. Here the sensing electrode consists of a small circular section inserted in a conducting ground plane. With an ideal current detector, the sensing electrode will also be at ground potential, and its charge will be given by

$$q_s = \int_{\theta=0}^{2\pi} \int_{r=0}^{a} \epsilon E_z r \, dr \, d\theta \qquad (3.3.3)$$

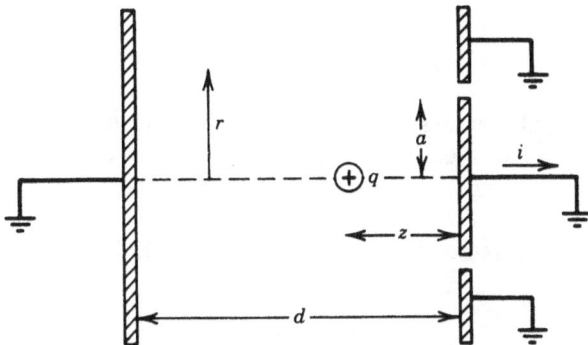

FIGURE 3.3.1.   A charge approaching a guarded electrode.

The electric field in this problem is not easy to solve. We simplify it still further by assuming that the charge is much closer to the right electrode than the left, so that substantially all of the electric field lines terminate on the right electrode. Under these conditions the electric field can be determined by the method of images, in which an equal and opposite charge is placed at the same distance from the other side of the surface, as shown in Figure 3.3.2. The electric field at the electrode surface is given by the superposition of the fields from the two charges as

$$E_z = \frac{-qz}{2\pi\epsilon(r^2 + z^2)^{3/2}} \tag{3.3.4}$$

Substituting this field into the charge Eq. (3.3.2) gives the charge on the sensing electrode as

$$q_s = -q\left[1 - \frac{z}{\sqrt{a^2 + z^2}}\right] \tag{3.3.5}$$

The current from the sensing electrode is given by Eq. (3.3.1) as

$$i = \frac{q}{a}\frac{dz}{dt}\frac{a^3}{(a^2 + z^2)^{3/2}} \tag{3.3.6}$$

This current will not be constant, even if the velocity is constant, since the position of the charge, which appears in the equation, is also changing. For constant velocity,

$$\frac{dz}{dt} = -U_0 \tag{3.3.7}$$

with the charge reaching the sensor at $t = 0$, the current expression is

$$i = \frac{qU_0}{a}\frac{a^3}{(a^2 + U_0^2t^2)^{3/2}} \tag{3.3.8}$$

which is plotted in Figure 3.3.3.

Discussion

The shape of this pulse, rising slowly to a broad peak and then rapidly dropping to zero, is characteristic of such charge sampling probes. If the sampler is directly in line with the approaching charge, this time of rise can be related to the speed of the charge. The height of the pulse depends on the magnitude of the charge, so a display of the current pulse allows us to determine both the charge and speed of a particle.

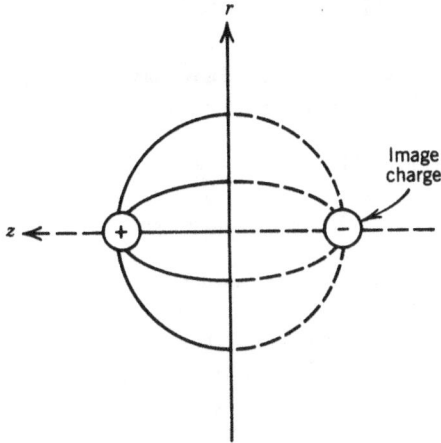

FIGURE 3.3.2.   Fields with an image charge.

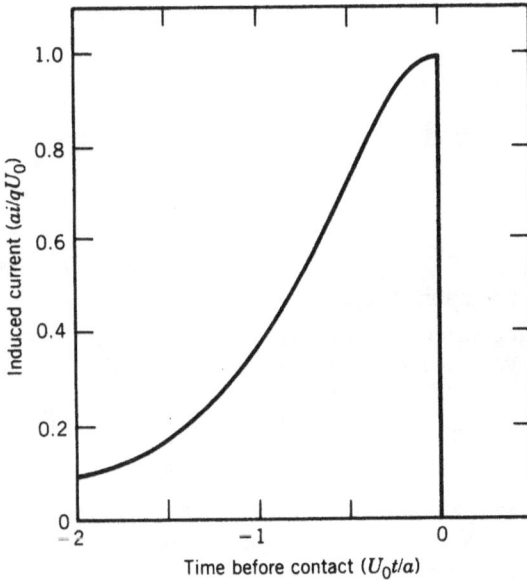

FIGURE 3.3.3.   Current induced by a charge approaching a guarded electrode.

The current has the same form whether the charge is transferred to the electrode or not. If the conductor is covered with an insulator (or blocking layer), the induced current will drop to zero as soon as the charge stops moving, whether its motion is stopped by the insulator or by the conducting electrode. Thus it is not possible to distinguish between a bare electrode or one covered by an insulator in terms of the external circuit current.

BIBLIOGRAPHY

Dessauer, J. H., and H. E. Clark, *Xerography and Related Processes*, Focal Press, London, 1965.

Eden, H. F., Electrostatic nuisances and hazards, in *Electrostatics and Its Applications*, A. D. Moore, Ed., Wiley, New York, 1973, pp. 425–440.

Hemenway, C. L., R. W. Henry, and M. Caulton, *Physical Electronics*, Wiley, New York, 1962, Ch. 7.

Klinkenberg, A., and J. L. van der Minne, *Electrostatics in the Petroleum Industry*, Elsevier, Amsterdam 1958, Ch. 8.

Moon, P., and D. E. Spencer, The poisson equation, in *Field Theory for Engineers*, Van Nostrand, Princeton, NJ, 1961, Ch. 14.

Ream, G. L., Servo control of multinozzle ink-jet operating point, *IEEE Trans. Ind. Appl.*, **IA-20**: 282–288 (1984).

Schaffert, R. M., *Electrophotography*, Focal Press, London, 1966.

Sessler, G. M., Ed., *Electrets*, vol. 1, Laplacian Press, Morgan Hill, CA, 1998.

Wilkinson, D. H., *Ionization Chambers and Counters*, Cambridge University Press, Cambridge, 1950.

PROBLEMS

PROBLEM 3.1   (ELECTRIC POWER)

One charge layer ρ_s is separated from an opposite charge layer by an air gap of length d (= 1 mm).

- a. If $V > V_{bk}$ and $E > E_{bk}$ must both be satisfied for breakdown, find the breakdown condition in terms of ρ_s.

- b. A conducting sheet of thickness a (= 0.5 mm) is inserted between the electrodes. What is the new breakdown condition?

PROBLEM 3.2   (MICROPHONE)

Charge can be carried around to be inserted into such devices as foil electret microphones by the following process. A plastic film (e.g., Teflon) is attached to a foil electrode on one side, while the other side is charged by an electron beam which coats this surface with a uniform charge layer ρ_s. The film has a thickness of a (= 25 μm) and a dielectric constant κ (= 2).

- a. What is the maximum charge that can be put on the Teflon before the breakdown field E_{bk} (= 10 MV/m) is exceeded? (The upper electrode, in the air, is very far away.)

- b. The upper grounded electrode is now brought to a distance b from the plastic. At what distance will the *air* breakdown field (3 MV/m) be exceeded?

- c. Will the voltage drop in the air exceed the 300 V minimum for breakdown at this distance?

PROBLEM 3.3 (PRECIPITATORS)

In an ES precipitator, charged dust builds up on the conducting wall in a layer of thickness a. As the dust layer gets thicker, breakdown may occur, causing a spark which blows the dust back into the exhaust air. Find the thickness of the layer when breakdown occurs ($E_{bk} = 3$ MV/m) in terms of the charge density ρ_0 ($= 10$ mC/m^3) inside the layer. Assume that the other electrode is very far away.

PROBLEM 3.4 (EXPLOSIONS)

A tall grain silo with an internal space charge density ρ_0 shows evidence of sparking, and you are asked if running a long wire down the axis of the silo will prevent sparking by lowering the maximum electric field inside. The silo is a cylinder of radius b; the wire has a radius a. Find the maximum E field with and without the wire and determine under what conditions the wire will help.

PROBLEM 3.5 (EXPLOSIONS)

A very wide oil tank of depth d ($= 20$ m) is partially filled with the oil to a depth a ($= 10$ m). The oil has acquired a space charge ρ_0 ($= 1$ mC/m^3) as a result of a pumping operation and has a dielectric constant ($\kappa = 2.5$).

  a. Find an expression for the electric field in the air above the oil.
  b. Does the field in the air increase or decrease as the level of the oil rises?

PROBLEM 3.6 (INSTRUMENTATION)

The spherical cap of a van der Graff generator is operated above breakdown so that corona occurs, charging the surrounding air with a charge density

$$\rho = \frac{\rho_0 a^4}{r^4}$$

where a ($= 0.1$ m) is the radius of the cap and ρ_0 ($= 1$ mC/m^3) is the charge density at its surface. If the cap is held at the voltage V ($= 1$ MV), what is the electric field at the surface ($r = a$)? Is this field greater than that obtained if there were no discharge (i.e., if $\rho_0 = 0$)?

PROBLEM 3.7 (ELECTRIC POWER)

The vertical electric field strength under ac transmission lines is often measured with a device consisting of a metallic sphere of radius a ($= 0.1$ m) separated electrically into an upper and a lower hemisphere. The external field induces charges on the hemispheres which vary with the external field, producing a current i. The relaxation time of air (~ 10 min) is much longer than the period of the ac field, so the air can be considered as insulating. Find the hemisphere current produced by the vertical field $E_z = E_0 \cos \omega t$ where E_0 (= 1 kV/m) is the external field strength and ω (= $2\pi 60$) is its frequency.

PROBLEM 3.8   (RADIATION DETECTORS)

In some radiation counters, such as the Geiger counter, charges of both signs are produced by ionizing radiation passing through the gas between two grounded electrodes. Some time after the radiation pulse, there are two uniform charge layers of density $\pm \rho_s$ (= 1 μC/m^2) moving in opposite directions toward the parallel electrodes, which are separated by a distance d (= 1 cm). If the charge layers are initially produced midway between the electrodes, and move with velocities $U_+ = U_0$ and $U_- = -2U_0$, sketch the current induced in each electrode as a function of time.

PROBLEM 3.9   (COMPUTER PERIPHERALS)

The IBM ink jet printer senses the timing of ink drop production (Ream, 1984) by measuring the charge induced by a droplet, passing parallel to a sensing electrode. This electrode, of width $2a$ (= 1 mm) is embedded in a guard electrode at a distance d (= 1 mm) at the same potential. Find the induced current in terms of the charge q (= 1 fC) and velocity (U_0 = 20 m/s) of the drop.

PROBLEM 3.10   (INSTRUMENTATION)

A wide electrode starts to emit a uniform charge cloud at $t = 0$. The cloud has a density ρ_0 (= 1 mC/m^3) and moves with a velocity U_0 toward a second electrode at a distance d (= 1 cm) from the first. Find the current i for $t > 0$.

4

PARTICLE MOTION IN KNOWN ELECTRIC FIELDS

In previous chapters we found the electric fields caused by charges with known locations. Although these charges may have been moving, their motion was given in advance and was not affected by electrostatic forces. In this chapter we begin to study more realistic situations in which the electrostatic forces affect or even dominate the motion of charged particles.

The simplest of these involves electric fields which are not in turn affected by the motion of the charges. Situations like this occur whenever there are relatively few charges present. In these cases the motion of the charged particle is given by solution of the equation

$$m\frac{d^2\mathbf{r}}{dt^2} = q\mathbf{E}(\mathbf{r}, t) + \mathbf{f}$$

$\mathbf{E}(\mathbf{r}, t)$ is a known function of space and time, whereas $\mathbf{f}$ represents all of the nonelectrical forces acting on the particle.

The solution of this equation has occupied scientists for many years, since it is the basic equation of classical mechanics. In this chapter we discuss several solutions which cover most of the cases of interest in practical devices based on electrostatic forces. At first we include only the electrostatic force. If this force depends only on time,

$$\mathbf{E} = \mathbf{E}(t)$$

we obtain a general solution, which is then used to describe the operation of a reflex klystron. If the electrostatic force depends only on space

$$\mathbf{E} = \mathbf{E}(\mathbf{r})$$

we again obtain a general solution involving potential energy concepts. This solution is applied to find the minimum size of an electron beam.

Drag forces are introduced in the last two sections. At first both inertia and drag are included, and a general solution involving a mechanical relaxation time is developed and applied to a calculation of drop trajectories in ink jet printers. Then inertia is completely neglected, giving the mobility limit, with an unusually simple general solution. This solution is used to calculate the charging rate for smoke particles in an electrostatic precipitator.

4.1   INERTIAL MOTION IN TIME-DEPENDENT FIELDS
(Reflex Klystrons)

Summary

The position of the charges is often affected by the electric forces exerted on the charge. The simplest possibility involves charges which experience only the electric force from a known electric field which is externally controlled. The only mechanical effect is the inertia of the charge. An example of this situation is the reflex klystron, which is used in many microwave and radar installations.

Theory

In the preceding chapter, the electric fields and voltages were determined under the assumption that the magnitude and location of all of the charges were known in advance. Actually, it is unusual for this information to be available, since the charges experience electrostatic forces in practically all applications, and they often move in response to these forces. As in any mechanical system, the effect of the applied force will be modified by various mechanical effects such as inertia, friction, and diffusion.

Often, however, only one of these effects is dominant in a particular application, so that the resulting motion of the charge is relatively easy to determine. For instance, if the charge is moving through a vacuum or is attached to a massive particle, inertia will dominate, and the position of the charge is described by the basic force balance equation,

$$m\frac{d\mathbf{u}}{dt} = q\mathbf{E}(t) \tag{4.1.1}$$

The electric field is uniform (constant in space), but it may vary in time. This equation is easy to solve by direct integration as

$$\mathbf{u} = \mathbf{u}_0 + \left(\frac{q}{m}\right)\int_0^t \mathbf{E}(t)\,dt \tag{4.1.2}$$

Once the velocity is known, the position of the charge is given by

$$\mathbf{x} = \mathbf{x}_0 + \int_0^t \mathbf{u}(t)\, dt \qquad (4.1.3)$$

Often the electric field is constant, so that

$$\mathbf{E}(t) = \mathbf{E}_0 \qquad (4.1.4)$$

In this special case

$$\mathbf{u} = \mathbf{u}_0 + \left(\frac{q}{m}\right)\mathbf{E}_0 t \qquad (4.1.5)$$

and

$$\mathbf{x} = \mathbf{x}_0 + \mathbf{u}_0 t + \frac{1}{2}\left(\frac{q}{m}\right)\mathbf{E}_0 t^2 \qquad (4.1.6)$$

Example: The Reflex Klystron

In many ballistics problems, the external electric fields are so high that the repulsive effects of space charge can be completely neglected. One example of such a device is the reflex klystron (Harman, 1961, Ch. 7), which is used to amplify and generate microwave power for radar and broadcasting transmitters. In a reflex klystron a beam of electrons is directed past two electrodes attached to the signal voltage. This voltage sets up alternating electric fields which first accelerate and then decelerate the electrons in accordance with the signal to be amplified. The electrons then drift into a region in which a static electric field repels them toward a collecting electrode, which is usually the same physical structure as the modulating electrode.

As a result of the velocity modulation, electrons which enter this reflex region take different periods of time to travel from the entrance to the collector, with the faster particles taking longer. If the operating conditions are arranged correctly, both the slowest particles and the fastest, which enter a half cycle later, can be made to reach the electrode at the same time, giving a very strong pulse at the output. The basis of this effect and the conditions necessary to achieve it are discussed here.

Application of Theory

The reflex region, where the beam is turned around, is sketched in Figure 4.1.1. Electrons enter the reflex region through an opening in the grounded electrode at $x = 0$ with an initial velocity which varies according to the entrance time, $u_0(t_0)$. In most cases this velocity modulation can be represented by the sinusoidal expression

$$u_0(t) = K_1 + K_2 \cos \omega t \qquad (4.1.7)$$

After entering the reflex region, each charge experiences a constant electric field set up by the opposite electrode, which is maintained at a potential V. This

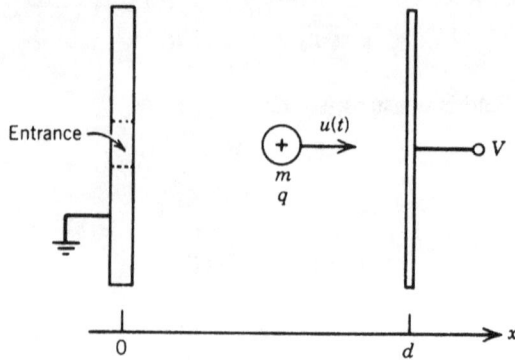

FIGURE 4.1.1. The reflex klystron.

electric field is given by

$$E_0 = -\frac{V}{d} \tag{4.1.8}$$

With a constant electric field, the position of the charge which enters the reflex region at t_0 can easily be obtained from Eq. (4.1.6) as

$$x(t) = -\frac{qE_0}{2m}(t - t_0)^2 + u_0(t_0)(t - t_0) \tag{4.1.9}$$

The time at which the charge strikes the collecting electrode can be obtained by setting the position given by this equation equal to the position of the collector, which in this case is $x = 0$. Thus the collection time is

$$t_c = t_0 + \frac{2mu_0(t_0)}{qE_0} \tag{4.1.10}$$

As expected, this time depends on the initial velocity of the electrons as they enter the reflex region and also on the applied voltage.

Discussion

Since the device operates by "bunching" the electrons at the collector, we would like to know the conditions needed for the fastest and slowest electrons to arrive at the collector at the same time. This implies that the time of collection is the same for both, or that

$$0 + \frac{2m}{qE_0}(K_1 + K_2) = \frac{T}{2} + \frac{2m}{qE_0}(K_1 - K_2) \tag{4.1.11}$$

where T is the period of the sinusoidal velocity modulation assumed in Eq. (4.1.7). This condition is satisfied at the frequency

$$f = \frac{1}{T} = \frac{qE_0}{8mK_2} \tag{4.1.12}$$

If the modulation amplitude is kept constant, the collector will give a bunched current output only at discrete frequencies, so that the device is frequency selective. Unlike many microwave devices, the resonant frequency is not determined solely by the dimensions of the reflex cavity, but also by the static applied voltage. Thus this device can be electrically tuned.

The bunching condition just presented only holds for the fastest and the slowest electrons. Those with intermediate speeds will not necessarily all reach the collector at the same time. When their trajectories are plotted using Eq. (4.1.9), however, it is clear that all of the trajectories tend to bunch around the same time as the fastest and the slowest, so that all of the particles are available to reinforce the initial modulation (Harmon, 1961). If the return time is chosen so that the returning electrons pass through the modulating grids when the electric field is opposing their momentum, they will be slowed down, and their kinetic energy will be transformed into electric energy in the source of the voltage, thus increasing the signal level. This effect is often used to make a microwave oscillator.

4.2   INERTIAL MOTION IN SPACE-DEPENDENT FIELDS
(Electron Beam Waist)

Summary

When the electrostatic field varies in space, the motion is described by a nonlinear differential equation which can usually be reduced to first order by a specific integrating factor related to the particle velocity. This reduction leads naturally to the idea of potential energy, while direct integration of the reduced equation gives the particle trajectory. The potential energy method is illustrated by calculating the minimum size of an electron beam, which limits the resolution in CRT terminals and electron beam lithography.

Theory

In Chapter 3, the electrostatic force on the charged particle was independent of particle position, although it could vary in time. A more difficult problem arises when the electric force varies in space, as it usually will in practice, where geometry and space charge cause field variations. Fortunately, if the force is independent of time, a general solution is still possible, based on the idea of the potential energy.

The equation for the particle takes the form

$$m\frac{d^2x}{dt^2} = qE(x) = f(x) \tag{4.2.1}$$

where x is any component of the position vector. The key aspects of the form here are the second derivative and the force which depends only on the unknown. If these are the only kinds of terms in the differential equation, it can always be solved by multiplying by an integrating factor dx/dt.

$$m \frac{dx}{dt} \frac{d^2x}{dt^2} = f(x) \frac{dx}{dt} \qquad (4.2.2)$$

It can be checked by differentiating that the first term is a perfect derivative

$$m \frac{dx}{dt} \frac{d^2x}{dt^2} = \frac{d}{dt} \left[\frac{1}{2} m \left(\frac{dx}{dt} \right)^2 \right] \qquad (4.2.3)$$

while the second term can always be written as

$$f(x) \frac{dx}{dt} = -\frac{\partial}{\partial x}[PE(x)] \frac{dx}{dt} = -\frac{d(PE)}{dt} \qquad (4.2.4)$$

$PE(x)$, which is called the potential energy, is just a function whose derivative gives the force. It can be determined in any problem by integrating the force,

$$PE(x) = -\int f(x) \, dx \qquad (4.2.5)$$

Thus the force equation takes the form

$$\frac{d}{dt} \left[\frac{1}{2} m \left(\frac{dx}{dt} \right)^2 + PE(x) \right] = 0 \qquad (4.2.6)$$

Now that the equation has been recast in the form of a perfect derivative, it becomes easy to integrate, so that

$$\frac{1}{2} m \left(\frac{dx}{dt} \right)^2 + PE(x) = TE \text{ (a constant)} \qquad (4.2.7)$$

This equation corresponds to the statement that the sum of the kinetic energy and the potential energy $PE(x)$ is a constant called the total energy TE. The value of TE is usually determined by knowing the potential energy (or position) and the kinetic energy (or velocity) at some initial time.

At this point, the problem has not been completely solved since we still have a first-order derivative, the velocity. In practice, however, the partial solution

$$\frac{dx}{dt} = \sqrt{\frac{2}{m}[TE - PE(x)]} \qquad (4.2.8)$$

is often all that is needed. If a particle initially is moving, it can continue moving throughout any region in which $dx/dt \neq 0$, or

$$PE(x) < TE \qquad (4.2.9)$$

Thus knowledge of $PE(x)$ and the initial energy determines the range of the particle as it moves in the so-called potential well.

If the actual position of the particle is needed as a function of time, it can be immediately obtained in integral form from Eq. (4.2.8) as

$$\int_0^x \frac{dx}{\sqrt{(2/m)[\text{TE} - \text{PE}(x)]}} = \int_0^t dt = t \tag{4.2.10}$$

Example: Electron Beam Size

The potential energy method finds wide application in any area in which charge motion is limited by inertia, as in solid crystals or in a vacuum. One common example is the electron beam, which finds such diverse uses as image display and integrated circuit fabrication. These beams are often more effective if the diameter is very small, but the beam size is limited by the mutual repulsion of the electrons as they are forced together. As a result, the beam usually has a *waist*, as shown in Figure 4.2.1. At the waist the initial inward velocity of the electrons is overcome by space charge repulsion, and the electrons are forced apart. The size of this waist can be determined by the potential energy method.

Application of the Theory

The size of the waist is determined by the radial equation of motion. In practice, the beam is much longer than its diameter and has an essentially constant axial velocity. Under these conditions the radial electric field depends only on the axial position, and the radial motion can be determined by following the position of an electron at the edge of the beam r,

$$m\frac{d^2r}{dt^2} = qE_r \tag{4.2.11}$$

Since all of the electrons in the beam are moving at a uniform axial velocity, the charge in any length of the beam is a constant ρ_1. The electric field at the edge of the beam is not constant, however, because the size of the beam varies. This field is given by the application of Gauss' law to the volume shown in Figure 4.2.2. as

$$E_r = \frac{\rho_L}{2\pi \epsilon r} \tag{4.2.12}$$

and the equation of motion for an electron at the edge of the beam becomes

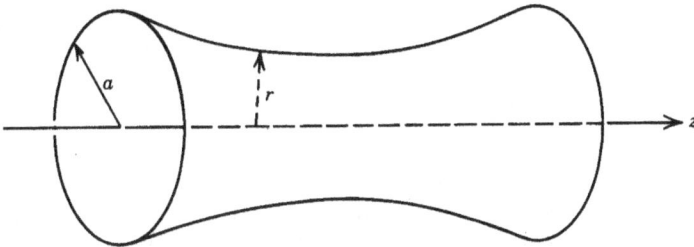

FIGURE 4.2.1. The waist of an electron beam.

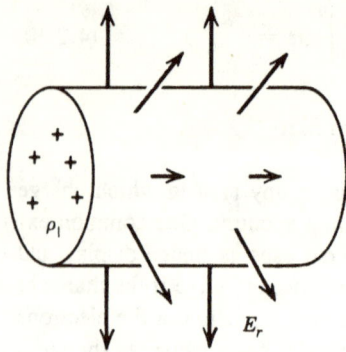

FIGURE 4.2.2. A short length of the beam.

$$m\frac{d^2r}{dt^2} = \frac{q\rho_L}{2\pi\epsilon r} = f(r) \tag{4.2.13}$$

This equation has a form suitable for the potential energy method. The potential energy is

$$PE(r) = -\int_a^r f(r)\,dr = \frac{q\rho_L}{2\pi\epsilon}\ln\left(\frac{a}{r}\right) \tag{4.2.14}$$

The reference position a can be chosen arbitrarily but should have the same value for all calculations of PE.

The total energy is given in terms of the initial inward velocity u_0 at radius a as

$$TE = \frac{1}{2}mu_0^2 - \frac{q\rho_L}{2\pi\epsilon}\ln\left(\frac{a}{a}\right) = \frac{1}{2}mu_0^2 \tag{4.2.15}$$

At the beam waist ($r = r_m$) the radial velocity is reduced to zero,

$$u(r_m) = 0 \tag{4.2.16}$$

so that

$$TE = 0 + \frac{q\rho_L}{2\pi\epsilon}\ln\left(\frac{a}{r_m}\right) \tag{4.2.17}$$

Equating the two expressions for TE given by Eqs. (4.2.15) and (4.2.17) yields

$$r_m = a\exp\left[\frac{-\pi\epsilon u_0^2}{(q/m)\rho_1}\right] \tag{4.2.18}$$

Discussion

As is often the case with the potential energy method, the first integral is all that is needed to obtain the desired answer, which in this case was the minimum beam size. In addition to electron beams, this method is also widely used in atomic and solid state physics to describe the motion of electrons. In these applications the primary result is also an allowed range of electron motion, expressed in terms of ionization energy or band energy. The potential energy method is also important in deriving $v–i$ relations for a variety of practical devices such as vacuum diodes and corotrons, discussed in later chapters.

Although the method has been presented here in terms of particle motion, it is, in essence, a basic integrating factor technique which can be applied to any second-order differential equation which lacks first derivatives and has no terms which depend explicitly on the independent variable. Poisson's equation often satisfies these conditions, so this method will also be useful when space charge influences the motion, as in Chapter 5.

4.3   DRAG FORCES VERSUS INERTIA
(Ink Jet Printers)

Summary

If charges move through some material medium such as a fluid or a solid, they will experience a drag, which tends to slow them down. This drag force is very important in practice, since most electrostatic devices do not operate in a vacuum. The effect of these forces is considered in this section in terms of an ink jet printer.

Theory

In many applications of electrostatics the charges are impeded in their motion by the friction or drag caused by the surrounding material. This can happen when charged drops move through air, as in an ink jet printer; when ions move through liquids, as in transformer oils; or when electrons move through solids, as in transistors. This friction, or drag, is one of the key parameters to be considered in designing electrostatic devices or in explaining electrostatic effects which occur naturally.

Although there are numerous effects which can give rise to this retarding force, most of them result in a force which is linearly proportional to the velocity, so that the equation of motion for the charge becomes

$$m\frac{d\mathbf{u}}{dt} = q\mathbf{E}(t) - K\mathbf{u} \tag{4.3.1}$$

This first-order equation has a general solution given by

$$\mathbf{u} = e^{-t/t_m}\int_0^t \frac{q\mathbf{E}(t)}{m}e^{t/t_m}\,dt + \mathbf{u}_0 e^{-t/t_m} \tag{4.3.2}$$

where $t_m = m/K$ and u_0 is the initial velocity. Of course the position follows from the velocity as

$$\mathbf{x} - \mathbf{x}_0 = \int_0^t \mathbf{u}(t)\,dt \tag{4.3.3}$$

The mechanical relaxation time t_m is the key parameter in determining how the charge will respond to external forces. If the force is applied for time which is very short compared to the mechanical relaxation time, the behavior will be inertial, and the analysis of Sections 4.1–4.2 will be appropriate. If the times of interest are much longer than the mechanical relaxation time, inertia will have little effect, and the particle will slow down under the influence of drag.

In practice, the electric fields are often constant and uniform

$$\mathbf{E}(t) = \mathbf{E}_0 \qquad (4.3.4)$$

so the solution simplifies to

$$\mathbf{u} = \frac{q\mathbf{E}_0}{K}(1 - e^{-t/t_m}) + \mathbf{u}_0 e^{-t/t_m} \qquad (4.3.5)$$

and

$$\mathbf{x} = \frac{q\mathbf{E}_0}{K}[t - t_m(1 - e^{-t/t_m})] + \mathbf{u}_0 t_m(1 - e^{-t/t_m}) + \mathbf{x}_0 \qquad (4.3.6)$$

Example: Ink Jet Printer

As an example of a situation where both the inertial and drag forces are important, we consider the motion of particles in a continuous ink jet printer sketched in Figure 4.3.1.

In this device, electrically conducting ink is forced through a nozzle to form a thin jet, which then breaks up into droplets under the influence of surface tension and the mechanical vibrations in the nozzle caused by a piezoelectric transducer. At the point where droplets are forming, an electric field controlled by a computer is applied to the jet to give the newly formed droplets an electric charge related to the signal. The drops then move into a deflection region where a steady transverse electric field deflects them by an amount which depends on the charge. This deflection causes them to strike the print surface (usually a piece of paper) at different points, creating an image. Since the droplets are very small, they can be

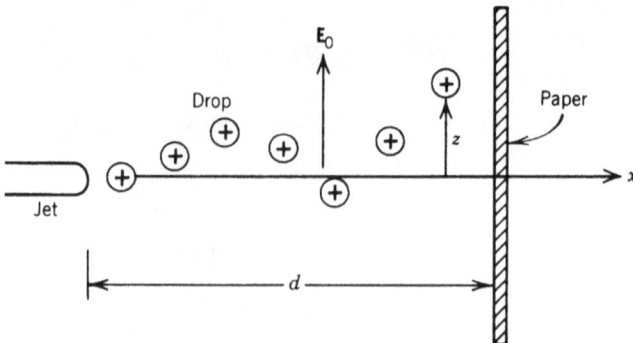

FIGURE 4.3.1. Ink jet printer.

quickly accelerated and are able to produce high quality hard copy much faster than electromechanical printers.

The quality of the image depends very strongly on the accuracy of placement of an individual droplet, and thus on its trajectory from the point of formation to the impact on the paper. There are two forces on the droplet during this time: the electric force and the drag force.

Application of Theory

Since the charged drops are widely separated in space and carry relatively small charges, the electric forces between droplets can be neglected in initial design work, and the only electric force which must be considered is that caused by the external uniform deflection field

$$\mathbf{f}_e = q\mathbf{E}_0 \tag{4.3.7}$$

The drag force is caused by the relative motion of the droplets through the air. This drag can be calculated in terms of the charge and constitution of the particle and is available for a wide range of shapes. For an ink drop, which can be modeled as a sphere of radius a with a density and viscosity much greater than the surrounding air, the viscous drag force is given by

$$\mathbf{F}_d = 6\pi\eta a\mathbf{u} = K\mathbf{u} \tag{4.3.8}$$

In this expression, η is the dynamic viscosity of the surrounding fluid, which varies in practice from 20 μPa-s for air to 1 Pa-s for thick oils. The size of the drop in an ink jet printer is determined by the resolution desired in the graphical output. In a typical printer the drop radius is given by $a \simeq 15\ \mu$m.

The transverse motion across the paper plane is strongly influenced by the static deflection field, so both electric and drag forces must be considered.

$$m\frac{du_z}{dt} = qE_0 - Ku_z \tag{4.3.9}$$

Since the ink drop has no initial velocity in the z direction, its velocity and displacement are given by Eqs. (4.3.5) and (4.3.6) as

$$u_z = \frac{qE_0}{K}(1 - e^{-t/t_m}) \tag{4.3.10}$$

$$z = \frac{qE_0}{K}[t - t_m(1 - e^{-t/t_m})] \tag{4.3.11}$$

A plot of the transverse velocity versus time is shown in Figure 4.3.2. The velocity initially increases with time but eventually reaches a limiting value. This occurs in a time given approximately by

$$t_m = \frac{m}{K} = \frac{m}{6\pi\eta a} = \frac{2}{9}\frac{\gamma a^2}{\eta} \simeq 2.5\ \text{ms} \tag{4.3.12}$$

where γ here is the density of ink ($\gamma \simeq 10^3$ kg/m^3). This is on the order of the

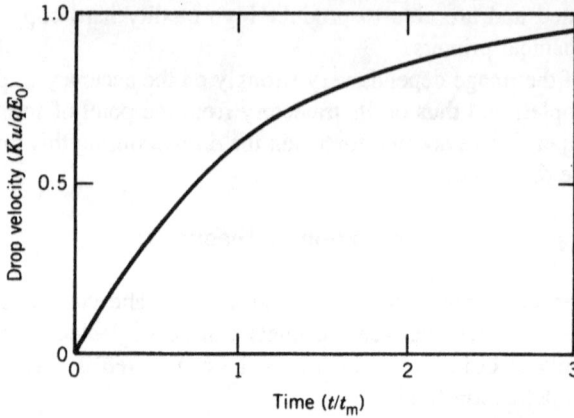

FIGURE 4.3.2. Velocity under the influence of a static electric field.

time that the drop takes to reach the paper, so both inertia and mobility are important here.

Discussion

Initially, the velocity increases linearly, just as it would in a vacuum. As the speed increases, however, more of the applied force must go toward overcoming drag, so that less is available for acceleration. Eventually the droplet reaches its terminal velocity, at which time the viscous drag is just balanced by the electrostatic force, giving the terminal velocity

$$U = \frac{qE}{K} = \mu E \qquad (4.3.13)$$

This is known as the mobility limit. Since it is extremely important for many practical applications of electrostatics, mobility is discussed in more detail in Section 4.4.

As we have demonstrated before, the maximum charge which can be applied to the drop is limited by the electric breakdown in air. For these drops, a reasonable value of charge might be $q \simeq 100$ fC, giving a mobility value of $\mu = 1.8 \times 10^{-5}$ m^2/V-s.

If drag effects could be completely eliminated ($\mu \to \infty$) inertia would still limit the drop motion to a maximum deflection at the paper of

$$z_m = \frac{(q/m)Et^2}{2} \simeq 16 \text{ mm} \qquad (4.3.14)$$

using the charge given previously, an electric field of 1 kV/mm, and an ink density of 10^3 kg/m^3. This is larger than the height of a typed character in most printing, so a single ink jet can be used to "paint" letters or any other desired shape under software control.

4.4   THE MOBILITY LIMIT
(Corona Charging of Particles)

Summary

When inertia can be neglected completely, the velocity is proportional to the local electric field, and the charge motion can be determined as soon as the field is known. This section presents an important application of this limit: the corona charging effect used in electrostatic precipitators to charge the dirt particles in smokestacks and to remove them from the exhaust gases.

Theory

When the mechanical relaxation time is short compared to the time which a charged particle spends in the electric field, the mobility limit is appropriate to describe the motion of particles. In the mobility limit the velocity of the ions is given by

$$\mathbf{u} = \mu\mathbf{E} \tag{4.4.1}$$

Thus the problem of finding charge motion reduces immediately to one of finding the field, covered in Chapters 2 and 3.

Example: Corona Charging in Electrostatic Precipitators

An example of such a situation is the bombardment (or corona) charging of particles in an electrostatic precipitator. These devices are widely used in industry to clean the exhaust gases from power plants, cement factories, and so on, and in home heating and air conditioning to remove suspended dust, which is often allergenic. Electrostatic precipation is also used with insulating liquids such as transformer oils to remove particles which may cause losses or breakdown. It is especially effective in removing particles which are too light to settle out under the influence of gravity alone.

In electrostatic precipitators, the particles pass through a region of the duct in which there is a copious supply of ions generated by a nearby corona discharge. Under the influence of an external electric field, the ions are forced across the duct, intercepting any particles which happen to be in the way. The charge that the particles acquire causes them to drift toward the walls where they are caught and eventually removed.

Naturally, a particle with a large charge will reach the collecting wall sooner than one with a small charge, giving more efficient precipitation. For this reason, designers try to ensure that each dirt particle collects as much charge as possible in the ionization region. Since the ion flow is governed by mobility, the maximum charge can be predicted in terms of the electrostatic fields which drive ions onto the particle's surface.

Application of the Theory

Solution of the electric field equations in the vicinity of a charged conducting particle is a common problem in applied electrostatics, so this example gives us the chance to apply a generally useful solution of the field equations. The particle will be taken as spherical, with a radius a (Fig. 4.4.1). The external field is assumed to be uniform, with a magnitude of E_0. At the time of interest the particle already has acquired a charge of magnitude q.

Although we have not solved this problem before, we already know the solution for two simpler related problems: the field around a charged sphere and the field around an uncharged sphere in a uniform field. Both of these solutions satisfy the same boundary conditions, except for the source of the field. Because Laplace's equation is linear, these two solutions can be added to give the solution for the combined problem, a charged sphere in an external field. The solution takes the form

$$\Phi = -E_0\left(r - \frac{a^3}{r^2}\right)\cos\theta + \frac{q}{4\pi\epsilon r} \tag{4.4.2}$$

$$E_r = E_0\left(1 + \frac{2a^3}{r^3}\right)\cos\theta + \frac{q}{4\pi\epsilon r^2} \tag{4.4.3}$$

$$E_\theta = -E_0\left(1 - \frac{a^3}{r^3}\right)\sin\theta \tag{4.4.4}$$

The electric field lines for this solution are shown in Figure 4.4.2 for different values of the particle charge. When the charge is zero, the field lines are symmetrically placed around the particle as shown in Figure 4.4.2a. As the particle collects charge, however, the lines avoid the particles owing to the repulsion by the charge already present, and the field pattern takes on the form of Figure 4.4.2b.

Now that the field pattern for the electric field is known, the flow of charge to the particle can be considered. It should be kept in mind that the charge is all of

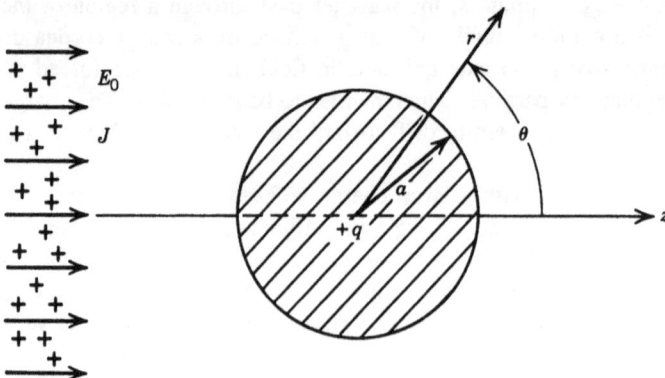

FIGURE 4.4.1. A particle under corona bombardment.

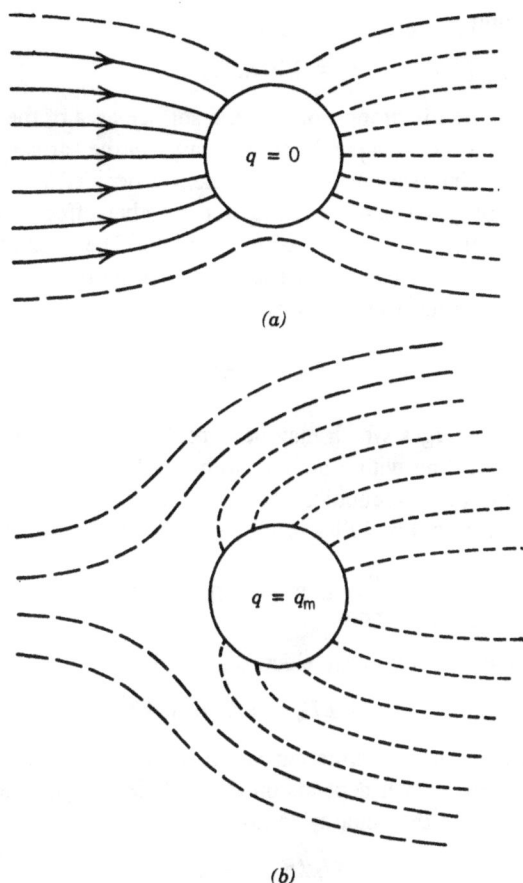

FIGURE 4.4.2. Electric fields in the vicinity of a charged particle. ——, lines which carry charge to particle; — —, lines which carry charge past particle; – – –, lines which carry no charge.

the same sign and that it has its source on the left, at large negative values of z. Thus only lines of electric force which originate to the left of the particle and intersect its surface can carry charge to the particle. The lines of electric field which leave the right side of the particle can not carry charge away because the particle can not act as a source of charge, that is, it does not emit ions as the corona source does. The angle θ_c is the smallest (or critical) angle at which charge arrives at the particle. This angle can be easily obtained from the solution of the electrostatic field problem, since it corresponds to the point at which the normal electric field goes to zero, or

$$E_r(r = a) = 3E_0 \cos \theta_c + \frac{q}{4\pi\epsilon a^2} = 0 \qquad (4.4.5)$$

The charge reaching the particle has only one sign, so the net charge on the particle must increase as time goes on, and the critical angle must therefore increase from its initial value of $\pi/2$ to its limiting value of π, when the particle charge reaches its limit

$$q_m = 12\pi\epsilon a^2 E_0 \tag{4.4.6}$$

The equilibrium charge is proportional to the surface area of the particle and to the applied electric field. This dependence of charge on the surface area is characteristic of electrostatic interactions and gives them a decided advantage in influencing the behavior of small particles compared to other effects, such as gravity and inertia which depend on the mass of the particle. The smaller a particle becomes, the larger is its ratio of area to mass, and the more effective the electric field becomes relative to most other forces.

Discussion

As an example of the charges which may be expected from this method, consider a typical precipitator dealing with a particle with a radius of 10 μm in an external electric field of 1 MV/m. The equilibrium electric charge is given by Eq. (4.4.6) as 33 fC, corresponding to a mobility in air [from Eq. (4.3.13)] of

$$\mu = \frac{q}{6\pi\eta a} = 8.9 \times 10^{-6} \text{ m}^2/\text{V-s} \tag{4.4.7}$$

which gives a particle velocity in the electric field of

$$u = \mu E_0 = 8.9 \text{ m/s} \tag{4.4.8}$$

The particle velocity in this situation is not a linear function of the applied field, since the mobility, which depends on the particle charge, will increase in a higher electric field. Thus the velocity follows the relation

$$u = \mu(E)E \sim E^2$$

As we demonstrate later, this nonlinear dependence on electric field violates the assumptions needed for Ohm's law and makes definitions of resistance ambiguous.

BIBLIOGRAPHY

Biblarz, O., *J. Electrostatics*, Vol. 5, 1978, p. 105.

Harmon, W. W., *Fundamentals of Electronic Motion*, McGraw-Hill, New York, 1961.

Ink Jet Printers, *IBM J. Res. Dev.*, **21**(1) 1–96 (Jan. 1977).

Pierce, J. R., *Theory and Design of Electron Beams*, Van Nostrand, Princeton, NJ, 1954.

Ralston, O. C., *Electrostatic Separation of Mixed Granular Solids*, Elsevier, Amsterdam, 1961.

Rice, S. A., and J. Jortner, The theory of ionic and electronic mobility in liquids, *Prog. Dielectr.*, **6**: 183–310 (1965).

Rose, H. E., and J. A. Wood, *Introduction to Electrostatic Precipitation in Theory and Practice*, 2nd ed., Constable, London, 1966.

Slater, J. C., and N. H. Frank, *Mechanics*, McGraw-Hill, New York, 1947, Ch. 1.

Sokolnikoff, I. S., and R. M. Redheffer, *Mathematics of Physics and Modern Engineering*, McGraw-Hill, New York, 1958, pp. 23–25.

Swatik, D. S., Nonimpact printing, in *Electrostatics and Its Applications*, A. D. Moore, Ed., Wiley, New York, 1973, pp. 307–335.

White, H. J., *Industrial Electrostatic Precipitation*, Addison Wesley, Reading, MA, 1963.

PROBLEMS

PROBLEM 1   (COMPUTER PERIPHERALS)

A CRT cathode at $x = 0$ emits a burst of electrons at zero velocity. They are accelerated toward a positive grid a distance a (= 1 cm) away held at a potential V_0 (= 1 kV). The electrons all pass through the grid and travel on in a field-free region until they strike the screen at a distance b (= 20 cm). Assume the electrons travel as a sheet of charge with surface density ρ_s (= 1 nC/m^2) and area A (= 10^{-4} m^2). Find and sketch as a function of time:

a.   The emitter current i_1

b.   The grid current i_2

c.   The screen current i_3

PROBLEM 2   (COMPUTER PERIPHERALS)

When using a CRT as a computer terminal, it is desirable to change the deflection of the electron beam from one spot on the tube face to another as quickly as possible. The high speed response can be studied in terms of a sinusoidal input voltage with a frequency ω. The deflection apparatus consists of two parallel electrodes of length d (= 2 cm) separated by a distance a (= 1 cm). The electrons enter the deflection region with a velocity U_0 (= 10^6 m/s) parallel to the electrodes, which are held at a potential difference of 100 V. Find the transverse velocity u_x at the exit and plot it versus frequency. Use this plot to estimate an upper rate at which the beam can be moved across the screen.

PROBLEM 3   (ASTRONAUTICS)

An ion rocket ejects a beam which is formed by emission of positive cesium ions from a wedge-shaped electrode (Harmon, 1961) into the region described by $x > |y|$. The potential field, neglecting space charge effects, is $\phi(x, y) = (V/a) \cdot (y^2 - x^2)$ where V (= 10 kV) is the accelerating voltage and a (= 10 cm) is a scaling length. The ions have single electronic charge q, a mass m (atomic weight = 133), and travel in a vacuum.

a.   Find the trajectory of an ion which leaves the electrode at $t = 0$ from the point x_0, y_0 with zero initial velocity.

b.   If the emission is confined to $-y_m < y < y_m$, what is the largest value of y which can be reached on any trajectory (i.e., the beam width)?

PROBLEM 4   (ELECTRIC POWER)

High pressure gas insulation (such as SF_6) is used in newer power substations to reduce the size of components. In the high electrical fields encountered in compact substations, a major leakage current is carried by small metal particles which bounce from one electrode to the other.

Consider a small hemispherical particle of aluminum, sitting on the lower electrode. The particle has a radius $a(= 0.5$ mm$)$ and a density γ $(= 2.7 \times 10^3$ kg/m$^3)$. The applied field induces a charge $q = -\frac{2}{3}\pi^3\epsilon_0 R^2 E_0$ on the particle.

a. What is the minimum field which could lift the particle?
b. What is the mechanical relaxation time and charge to mass ratio. (Use $\eta = 100$ μPa-s).
c. If the field is turned on to $E_0 = 3$ MV/m at $t = 0$, find the trajectory of the charged particle.

PROBLEM 5   (MINING)

In Florida, phosphate (for fertilizer) is separated from sand electrostatically. The ore passes through a hopper where the two types of particles are charged by friction to opposite polarities with a charge to mass ratio of q/m $(= 10$ μC/kg$)$. The particles have a radius a $(= 100$ μm$)$ and a density γ $(= 2 \times 10^3$ kg/m$^3)$. After charging, they fall under gravity between two electrodes which apply E_0 $(= 0.5$ MV/m$)$. The electrodes have a length d $(= 0.2$ m$)$.

a. Which type of motion is more likely here, mobility or inertia limited?
b. How far will the particles be separated when they leave the electrodes?

PROBLEM 6   (INSULATORS)

When positive and negative ions both occur in an insulator with dielectric constant κ $(= 3)$ their effect is reduced because opposite ions attract each other and recombine.

a. If two ions with opposite charge q_+ $(= 1.6 \times 10^{-19}$ C$)$ and q_- $(= -1.6 \times 10^{-19}$ C$)$ and mobilities μ_+ $(= 10^{-8}$ m^2/V-s$)$ and μ_- $(= 0.5 \times 10^{-8}$ m^2/V-s$)$ are initially separated by a distance d, how long will it take them to recombine?
b. If the density of negative ions is n_- $(= 10^{15}/$m$^3)$ estimate the distance d.
c. If there are n_+ $(= 0.5 \times 10^{12}$ m$^{-3})$ positive ions, how many recombinations take place in 1 s within a volume of 1 m^3?

5

UNIPOLAR SPACE-CHARGE MOTION

When large numbers of charged particles are present, their mutual repulsion strongly affects their motion so that the solutions of Chapter 4 are no longer useful. Because of the space charge, the electric fields are not known in advance and must be found by simultaneous solution of the mechanical and electrostatic equations. In this chapter we consider only situations in which there is a single type of charge carrier. In addition, generation and recombination are absent in the region under consideration, so that only motion of the charged particles can change the space-charge density. This situation often arises in vacuum or in materials at room temperatures.

We begin with the relatively simple mobility limit for mechanical motion and the assumption that the electrostatic field is due solely to the space charge. When the charge distribution is initially uniform, these assumptions give a simple charge decay, which describes the situation in oil tanks after a filling operation. Next, external fields are added to the space-charge fields to find the terminal relation for the corotron used in xerographic copiers. Finally, we consider the inertial limit of motion, which corresponds to Child's law. This result is applied to the design of an ion rocket engine.

5.1 MOBILITY AND SPACE CHARGE
(Unipolar Charge Decay in Tanks)

Summary

When externally applied fields are weak or absent, the only electrical force acting on the particle comes from the space charge of the particles themselves. This is the situation in a conducting fuel tank in which the fuel is charged in the filling operation. Decay of this charge is needed to prevent explosions, and the decay is caused solely by self-repulsion of the charges.

Theory

The decay of charge inside a volume which holds charges of a single sign can be easily studied by means of the basic equations of conservation of charge,

$$\nabla \cdot \mathbf{J} + \frac{\partial \rho}{\partial t} = 0 \tag{5.1.1}$$

Gauss' law,

$$\nabla \cdot \mathbf{D} = \rho \tag{5.1.2}$$

and the definition of current in a medium with charges of a single sign, all moving at the same velocity,

$$\mathbf{J} = \rho \mathbf{u} \tag{5.1.3}$$

Substituting the definition of current into conservation of charge gives

$$\left(\frac{\partial}{\partial t} + \mathbf{u} \cdot \nabla \right) \rho + \rho (\nabla \cdot \mathbf{u}) = 0 \tag{5.1.4}$$

Often the charge is uniformly distributed over the volume so that $\nabla \rho = 0$. Also, we will assume that the charges are moving at the mobility limit,

$$\mathbf{u} = \mu \mathbf{E} \tag{5.1.5}$$

and that the mobility is constant. This assumption is usually satisfied in liquids and solids, as well as in many gaseous media. With these assumptions, the last term in Eq. (5.1.4) can be written as

$$\rho (\nabla \cdot \mathbf{u}) = \rho \mu (\nabla \cdot \mathbf{E}) = \frac{\mu \rho^2}{\epsilon} \tag{5.1.6}$$

With this substitution, Eq. (5.1.4) becomes

$$\frac{\partial \rho}{\partial t} + \frac{\mu \rho^2}{\epsilon} = 0 \tag{5.1.7}$$

which has the solution

$$\rho = \frac{\rho_0}{1 + \mu \rho_0 t / \epsilon} \tag{5.1.8}$$

where ρ_0 is the initial charge density at $t = 0$.

As expected, this charge decreases as time goes on, with a significant decrease occurring in a characteristic charge decay time of

$$t_e = \frac{\epsilon}{\mu \rho_0} \tag{5.1.9}$$

The decay time depends on the initial charge density, so that large charge densities decay faster than small ones.

After several decay times, the charge density of Eq. (5.1.8) reaches a limiting relation of

$$\rho \longrightarrow \frac{\epsilon}{\mu t} \tag{5.1.10}$$

which is independent of the initial charge density. This is a surprising result, since it states that no matter how much charge was initially present, the amount remaining after a given elapsed time can not be larger than an amount which depends only on material properties and the elapsed time.

Example: Charge Decay in an Oil Tank

Whenever there is relative motion between two materials, charges are generated at the point of contact. If the materials are good insulators, these charges may build up to generate high voltages, occasionally causing electric discharges. These effects are well known with solid materials and lead to unpleasant shocks throughout the winter in northern climates. Although less well known, there is a dangerous parallel to this effect when insulating liquids flow past a solid surface. The charge that is built up by the flow of liquids such as petroleum can reach considerable proportions and has on some occasions led to serious tanker explosions initiated by filling and cleaning operations. When a tank is filled with an insulating liquid, the charges introduced by the flowing petroleum can be dissipated only by migration to the walls, and until this process is completed there is a potential explosion hazard.

Application of the Theory

As an example, consider a liquified natural gas (LNG) tank with a charge density of $\rho_0 = 1$ nC/m^3, a mobility of $\mu = 10^{-8}$ m^2/V-s, and a dielectric constant $\kappa = 2$. The charge decay time is given by Eq. (5.19) as 1.8×10^6 s, which represents a time on the order of months. Even if the charge density were much higher, say 1 μC/m^3, the decay time would still be 1.8×10^3 s or approximately 30 min. These figures suggest that once a charge is introduced into an insulating material, it may take quite a while for them to leave under their own repulsion.

Discussion

It should be noted that the charge density does not decay exponentially, which is the usual situation in electric circuits. Exponential decay can be expected only when specific constraints on mobility and charge density are met. These constraints are often the same as those needed to use Ohm's law to describe the medium. Since Ohm's law is rarely valid in the highly insulating materials which form the heart of most electrostatic devices, exponential decay is not often encountered. The difference between the two decay processes is illustrated in

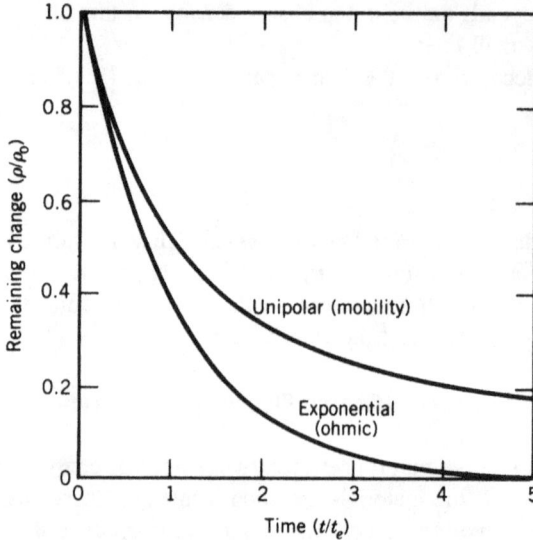

FIGURE 5.1.1. Charge decay in mobility-limited and ohmic material.

Figure 5.1.1. Although the initial decay rates are identical, the amount of charge remaining after some time is always much greater in mobility-dominated materials. These materials are therefore harder to discharge and more likely to be affected by triboelectric effects.

5.2 MOBILITY, SPACE CHARGE, AND EXTERNAL FIELDS
(Xerographic Corotrons)

Summary

In this section we consider the behavior of charges in a mobility-dominated case when both the external electric fields and the space-charge repulsion of the particles themselves are important. This situation often arises in practical devices which need high unipolar charge levels, such as xerographic copiers.

Theory

In the last section, the space-charge density was large enough to account for most of the dynamics of the charges and led to its own decay in time. The space charge can also have important effects when there are external fields applied to the system. As in any unipolar space-charge problem, the equation of conservation of charge

$$\frac{\partial \rho}{\partial t} + \nabla \cdot \mathbf{J} = 0 \qquad (5.2.1)$$

plays a key role. This form of the equation already neglects the generation or loss of charge, and we will make the additional assumption of steady-state behavior

$$\frac{\partial \rho}{\partial t} = \nabla \cdot \mathbf{J} = 0 \tag{5.2.2}$$

Gauss' law

$$\nabla \cdot \mathbf{E} = \rho/\epsilon \tag{5.2.3}$$

is also needed in a space-charge problem. Finally, the expression for current in the mobility limit is

$$\mathbf{J} = \rho \mathbf{u} = \mu \rho \mathbf{E} \tag{5.2.4}$$

These equations appear simpler than those governing charge decay, which were given in the preceding section. In steady-state current flow, however, charge is continuously supplied by an external source through electrodes, and the charge density varies throughout the volume. A general equation can be developed for this situation by substituting Eqs. (5.2.3) and (5.2.4) into conservation of charge [Eq. (5.2.2)], which yields

$$\nabla \cdot [\mathbf{E}(\nabla \cdot \mathbf{E})] = 0 \tag{5.2.5}$$

if mobility and permittivity are constants. This second-order equation will require two boundary conditions for a solution. One of these conditions is usually related to the total voltage drop between the electrodes which supply the charge,

$$V = -\int_{-}^{+} \mathbf{E} \cdot d\mathbf{r} \tag{5.2.6}$$

whereas the other involves the electric field which must be present at the emitting electrode to supply the required charge flow,

$$\mathbf{E}_0 = \mathbf{E}_0(\mathbf{J}) \tag{5.2.7}$$

In many cases the principal limitation to the current comes from the repulsion of the charges already present between the electrodes and not from the emission limits of the electrode itself. If so, only a very small electric field is needed at the emitting electrode, and the boundary condition there becomes

$$\mathbf{E}_0 \simeq 0 \tag{5.2.8}$$

This is known as the space-charge limit.

Example: The Corotron

Xerographic copying machines require large amounts of unipolar charge to prepare the photoreceptor to receive an image and to aid in removing the image before a succeeding copy is made. This charge is usually supplied by a corotron, consisting of a fine wire held at a high enough voltage to cause corona breakdown. The discharge at the wire supplies copious amounts of unipolar charge,

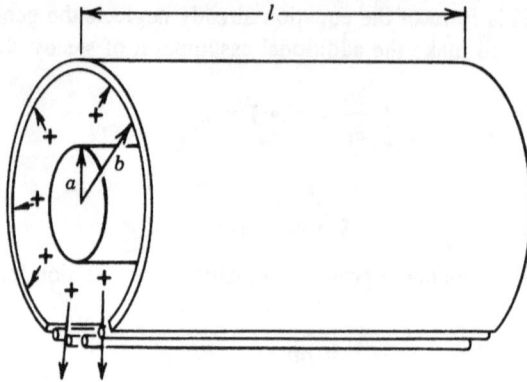

FIGURE 5.2.1. A corotron.

which is then drawn to the surrounding electrode by electrostatic forces, as shown in Figure 5.2.1. There is an opening in part of the outer electrode which allows the charge to escape. This charge flow, which is then used in the copier to prepare the photoreceptor, can be predicted by solution of the equations just given.

Application of the Theory

In solving the space-charge flow equations, it is often simpler to attack the charge conservation and Poisson's equations successively, rather than combining them as in Eq. (5.2.5). Beginning with

$$\nabla \cdot \mathbf{J} = \frac{1}{r} \frac{\partial}{\partial r} (rJ_r) = 0 \tag{5.2.9}$$

in cylindrical coordinates, we find

$$J_r = \frac{K}{r} \tag{5.2.10}$$

The constant K can be related to the total current at an electrode. At the emitting wire ($r = a$), the total current is

$$i = \iint \mathbf{J} \cdot d\mathbf{S} = \int_{z=0}^{l} \int_{\theta=0}^{2\pi} \frac{K}{a} (a\, d\theta\, dz) \tag{5.2.11}$$

which gives the current density inside the corotron as

$$J_r = \frac{i}{2\pi lr} \tag{5.2.12}$$

The charge density is given by Eq. (5.2.4) as

$$\rho = \frac{J_r}{\mu E} = \frac{i}{2\pi \mu lr E}$$

Substitution of this expression into Gauss' law, Eq. (5.2.3), gives the last equation to be solved,

$$\frac{1}{r}\frac{\partial}{\partial r}(rE) = \frac{i}{2\pi\mu\epsilon lrE} \tag{5.2.13}$$

in cylindrical coordinates. Multiplying both sides by r^2E and integrating from the wire out gives

$$E^2 = \left(\frac{a}{r}\right)^2 E_0^2 + \frac{i}{2\pi\mu\epsilon l}\left(1 - \frac{a^2}{r^2}\right) \tag{5.2.14}$$

where E_0 is the electric field needed to maintain the current flow at the emitting wire.

In a corotron the wire is usually much smaller than the diameter of the electrode. (Typical dimensions are 0.05 mm versus 10 mm). Except for a small region very close to the wire, $(a/r)^2 \ll 1$, and the electric field throughout the device is approximately uniform. The total voltage drop under these conditions is given by

$$V = \int_a^b \mathbf{E} \cdot d\mathbf{r} \simeq \left(\frac{b^2 i}{2\pi\mu\epsilon l}\right)^{1/2} \tag{5.2.15}$$

(if $E_0 = 0$) which gives the terminal relation of the corotron as

$$i = 2\pi\mu\epsilon l\left(\frac{V}{b}\right)^2 \tag{5.2.16}$$

Discussion

This approximate corotron terminal relation (Fig. 5.2.2) is only valid if the applied

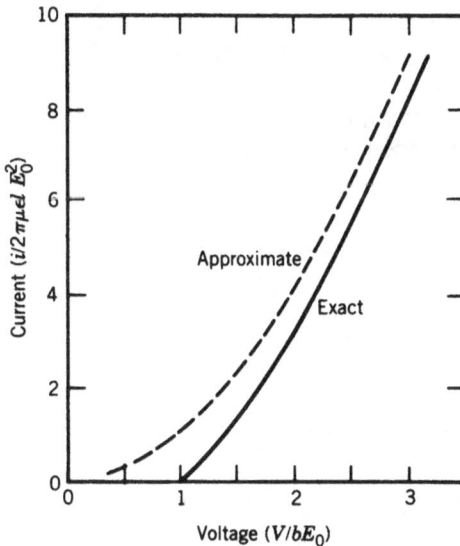

FIGURE 5.2.2.   Terminal relation for a corotron.

voltage is very large,

$$V \gg bE_0 \qquad (5.2.17)$$

At lower voltages the electric field at the wire may not reach the value needed to emit electrons, which reduces or cuts off the current completely. The effect of including E_0, shown in the figure, is to reduce the current at every voltage.

The quadratic dependence on voltage is typical of mobility flow limited only by space charge and occurs in any geometry, not just the cylindrical case considered here. It is often seen in fluids and also in solid-state devices (Shur and Eastman, 1979). It should be noted that this device does not satisfy Ohm's law, which would require a linear dependence on the applied voltage.

5.3   INERTIAL MOTION WITH SPACE-CHARGE FIELDS
(Child's Law and Ion Rockets)

Summary

In many applications, especially where carriers of one sign predominate, high currents can be obtained only with high space charges. If the space charge is large enough to distort the applied electric field, it can have a great effect on the operation of the device. In addition, since the space-charge field has an important effect on the motion of the charges, both the motion and field must be determined simultaneously. A classical example of this situation is Child's law, which describes current flow through a vacuum between an emitting and an absorbing electrode. Although initially formulated to describe a vacuum diode, Child's law is now useful in many additional areas, such as semiconductor devices at low temperatures and ion rockets for deep space exploration.

Theory

Just as in the previous section, the space-charge flow must satisfy conservation of charge and Poisson's equation. In the steady state

$$\nabla \cdot \mathbf{J} = 0 \qquad (5.3.1)$$

and with constant permittivity

$$\nabla \cdot \mathbf{E} = \rho/\epsilon \qquad (5.3.2)$$

The current density is again given by

$$\mathbf{J} = \rho\mathbf{u} \qquad (5.3.3)$$

but when the motion is dominated by inertia the expression for the velocity is not as simple as it was for mobility-dominated motion.

The motion of each charge is given by Newton's law

$$m\frac{d^2\mathbf{r}}{dt^2} = q\mathbf{E}(\mathbf{r}) \qquad (5.3.4)$$

Fortunately the electrostatic force is conservative, so potential energy techniques can be used, as described in Section 4.2. Thus the velocity of the charge at any point is given by

$$\mathbf{u(r)} = \mathbf{i}_u \sqrt{\frac{2}{m}[TE - q\Phi(\mathbf{r})]} \tag{5.3.5}$$

where $\mathbf{i}_u$ is a unit vector in the direction of motion. Substituting this expression for velocity along with Gauss' law into conservation of charge [Eq. (5.3.1)] gives a single equation for the potential as

$$\nabla \cdot [\mathbf{i}_u \sqrt{(TE - q\Phi)} \, \nabla^2\Phi] = 0 \tag{5.3.6}$$

with constant permittivity.

This is a third-order equation which requires three boundary conditions. Two of them are similar to those used in mobility-dominated space-charge flow; the total voltage drop is related to the internal electric field

$$V = -\int_-^+ \mathbf{E} \cdot d\mathbf{r} \tag{5.3.7}$$

and the field at the emitting electrode is sufficient to supply the needed current density

$$\mathbf{E}_0 = \mathbf{E}_0(\mathbf{J}) \tag{5.3.8}$$

The third condition, which is related to the introduction of inertia, involves the initial velocity of the charge carrier as it leaves the electrode. Often this initial velocity is assumed to be negligibly small.

Example: Ion Rocket Engine

There are many devices in which the space-charge density is very high, and the motion of the charges is influenced to a large degree by the repulsion of the surrounding charges. One example of such a device is an ion rocket engine (Jahn, 1968) consisting of a heated anode which emits ions toward a cool cathode some distance away in a vacuum, as shown in Figure 5.3.1.

If a negative voltage is placed on the cathode, ions will be attracted, causing a current flow. At the cathode, or neutralizing grid, which is constructed with large openings, low energy electrons are added to the ion beam to neutralize it, and both ions and electrons then continue through the openings in the cathode out the back of the rocket. This flow of particles leads to a thrust in the opposite direction, which propels the rocket forward.

Since the particles involved are very light and can be accelerated to very high velocities, the rocket can be operated for long periods of time on very little fuel. If a spacecraft is to remain operational for several years, such as a communications satellite, the fuel must be used very sparingly. For this reason ion rockets are often selected for this mission. [These rockets were also chosen for the TIE (Twin Ion Engine) fighters of Star Wars' fame.] The power produced by the engine is pro-

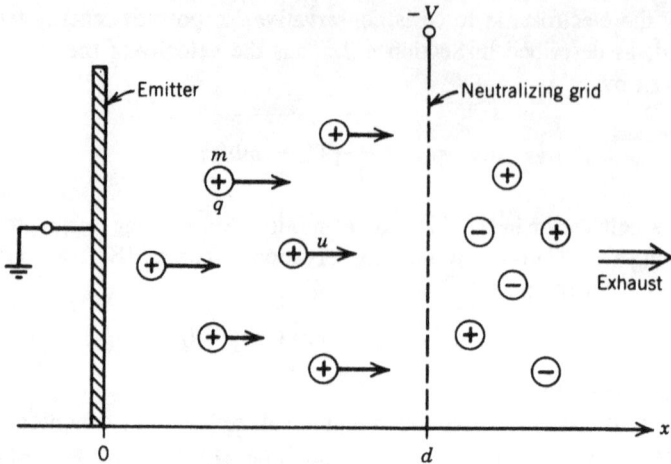

FIGURE 5.3.1. An ion rocket.

portional to the current output, so the power is limited by the same space-charge effects responsible for the current limitation.

Application of Theory

In applying the theory to a practical analysis, it is usually simpler to solve conservation of charge first, and then Poisson's equation. Conservation of charge in rectangular geometry takes the form

$$\frac{dJ}{dx} = 0 \tag{5.3.9}$$

or

$$J = J_0 \tag{5.3.10}$$

which gives

$$\rho = \frac{J_0}{u} \tag{5.3.11}$$

from the definition of current density. The charge velocity is given by the potential energy method as

$$u = \sqrt{\frac{2}{m}(\text{TE} - q\Phi)} \tag{5.3.12}$$

The constant TE is determined from the initial velocity of the ion, since

$$\text{TE} = \tfrac{1}{2}mu_0^2 + q\Phi(x = 0) \tag{5.3.13}$$

In this application $\Phi(x = 0) = 0$, and the emission velocity of the ion is usually very slow compared to its exhaust velocity, so $u_0 \simeq 0$. Thus

$$TE = 0 \tag{5.3.14}$$

and the velocity becomes

$$u = \sqrt{-2(q/m)\Phi} \tag{5.3.15}$$

Poisson's equation then becomes

$$\frac{d^2\Phi}{dx^2} = \frac{-J_0(-\Phi)^{-1/2}}{\epsilon\sqrt{2(q/m)}} \tag{5.3.16}$$

Although this equation is nonlinear, it has the same form as Newton's law, namely, a second derivative equal to a function of the unknown. Just as in the potential energy method, this equation may be solved by multiplying both sides by an integrating factor, which in this case is $d\Phi/dx$, and integrating to get

$$E^2 = \left(-\frac{d\Phi}{dx}\right)^2 = \frac{4J_0(-\Phi)^{1/2}}{\epsilon\sqrt{2(q/m)}} + E_0^2 \tag{5.3.17}$$

Since the potential is taken as zero at the anode ($x = 0$), the constant E_0 is related to the electric field there. This electric field will arise whenever the ions need an electric force to remove them from the anode at a rate sufficient to supply the current flow. Often the assumption is made that the electric field at the anode emitter is very small, since it simplifies the solution of the equation. This implies that the emitter is capable of supplying (and absorbing) large numbers of ions, which may not always be true at the desired current level. If the assumption is true, the anode is termed an *ohmic contact*, and the current has the largest possible value for that arrangement. In the case of an ion rocket, it is not always possible to supply all the needed ions, but the ohmic contact limit represents a goal which we try to reach by good design.

In this section we assume that this goal has been reached and that the emitter does not limit the current, so that

$$E_0 = E(x = 0) = 0 \tag{5.3.18}$$

Solution of Eq. (5.3.17) gives the potential as

$$\Phi(x) = -V\left(\frac{x}{d}\right)^{4/3} \tag{5.3.19}$$

the electric field as

$$E(x) = \frac{4}{3}\frac{V}{d}\left(\frac{x}{d}\right)^{1/3} \tag{5.3.20}$$

and the charge density in the gap as

$$\rho(x) = \epsilon\frac{dE}{dx} = \frac{4}{9}\frac{\epsilon V}{d^2}\left(\frac{d}{x}\right)^{2/3} \tag{5.3.21}$$

Equation (5.3.19) can also be used to give the output characteristics of the ion rocket by evaluating the current and voltage at the neutralizing grid ($x = d$). This procedure gives the current–voltage relation as

$$J_0 = \rho u = \frac{4}{9} \frac{\epsilon \sqrt{2q/m}}{d^2} V^{3/2} \qquad (5.3.22)$$

This is known as Child's law for a vacuum. The current given by Child's law is the maximum current that can be obtained from a device which depends on the flow of unipolar charge in a vacuum and thus sets the upper limit (the *space-charge limit*) for the output.

In a typical design for an ion rocket, the electric field is limited by the breakdown strength of vacuum to values below 10 MV/m. To keep the voltages as low as possible, the spacing between the electrodes should be small, but since it is difficult to maintain a fixed spacing over a large width in space, a compromise value of $d = 5$ mm is selected. This requires a voltage of 37.5 kV. Cesium is often used in these rockets because it is easy to ionize; Cs ions have a charge to mass ratio of $q/m = 7.25 \times 10^5$ C/kg and will acquire an exhaust velocity of $v = 2.3 \times 10^5$ m/s when accelerated through the electrodes.

Discussion

Devices limited by space charge have a variable electric field in the gap (Fig. 5.3.2a) which is much less than the equivalent uniform field V/d at the emitter and much greater at the opposite electrode. This high field makes breakdown more likely and therefore limits the operating voltage. Since the exhaust velocity, and hence the thrust, of the engine increases with the applied voltage, space-charge effects pose a serious limitation on performance by reducing the allowed operating voltage as well as the current. Vacuum tubes are also limited in this way, but transistors are not since they use carriers of both signs enabling them to neutralize

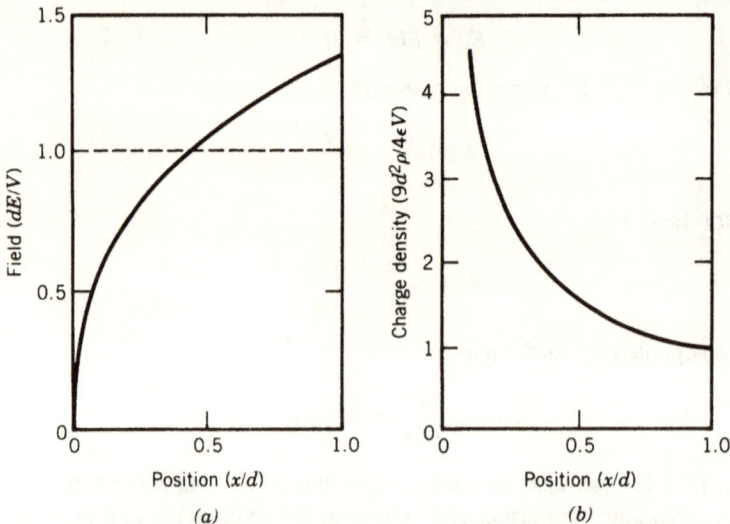

FIGURE 5.3.2. Field and charge distribution between electrodes.

space charge and thus carry larger currents. This is one of the reasons why transistors have supplanted vacuum tubes in most high current applications.

The plot of charge density (Fig. 5.3.2b) indicates that the density is infinite at the emitter. This results from our initial assumption that the emitter could supply as many charges as needed without the need for a large electric field, together with the assumption that the charges are stationary ($u = 0$) at the emitter. In practice neither of these assumptions is strictly valid, so the charge density, although very high at the emitter, is still finite.

BIBLIOGRAPHY

Birdsall, C. K., and W. B. Bridges, *Electron Dynamics of Diode Regions*, Academic, New York, 1966.

Cobine, J. D., *Gaseous Conductors*, Dover, New York, 1958.

Harman, W. M., *Fundamentals of Electronic Motion*, McGraw-Hill, New York, 1953, Ch. 5.

Jahn, R. G., *Physics of Electric Propulsion*, McGraw-Hill, New York, 1968, Ch. 7.

Klinkenberg, A., and J. L. van der Minne, *Electrostatics in the Petroleum Industry*, Elsevier, Amsterdam, 1958.

Lampert, M., and P. Mark, *Current Injection in Solids*, Academic, New York, 1970.

Shur, M. S., and L. F. Eastman, Ballistic transport in semiconductors at low temperatures for low-power high-speed logic, *IEEE Trans. Electron. Devices*, **ED-26:** 1677–1683 (1979).

PROBLEMS

PROBLEM 1   (INSTRUMENTATION)

A unipolar smoke detector consists of two parallel electrodes, separated by a distance d (= 1 cm), with a steady applied voltage V (= 100 V). One electrode contains a radioactive material which ionizes the air nearby, forming ions of both signs. One polarity is immediately absorbed by the electrode, and the other polarity travels to the opposite electrode at a speed $u_i = \mu_i E$. The charges are supplied in such abundance that the current is limited only by the space charge that builds up in the air between the electrodes.

a. Find the space-charge-limited current.

b. When smoke particles enter the space, the ions stick to them so that the charge now moves with the mobility of a charged smoke particle, which is $\mu_s \simeq 0.001 \, \mu_i$. What effect will this have on the current i?

PROBLEM 2   (ELECTRONICS)

A capacitor in an integrated circuit is formed by growing a SiO_2 layer of thickness d (= 1 μm) over the conducting silicon substrate and covering it with a metal electrode of area A.

a. If the capacitance must have the value $C = 10^{-9}$ F, what is A?

b. A dc voltage V (= 10 V) is applied to the capacitor. Assume that all con-
    duction is due to electrons injected from the silicon and the interface at the
    bottom is ohmic. Assume that the electron motion is mobility limited, with
    a mobility μ (= 0.1 m^2/V-s) through the SiO$_2$ which has a dielectric con-
    stant of κ (=4). What is the leakage current? (The answer will be much
    larger than typical measured currents, indicating that space charge effects
    can be neglected in practice)
c. Sketch the potential and E field distribution in the SiO$_2$ layer.

PROBLEM 3   (ELECTRONICS)

Find the relation between current and voltage for parallel electrodes with mobility-
limited conduction when the injecting contact is nonohmic. Assume that E_0 is a
constant.

PROBLEM 4   (ELECTRONICS)

Find the relation between current and voltage for a cylindrical vacuum diode with
the inner, emitting electrode of radius a and the outer electrode of radius b.

6

CHARGED PARTICLE CONSERVATION

The electric field affects the behavior of charged particles in every application, but additional processes can also influence the charges and, in some cases, may even dominate. Four of these additional influences (generation, recombination, convection, and diffusion) are covered in this chapter.

All four of these processes involve a change in the number of charges in a given region. They are conveniently grouped under the heading of particle conservation, since a single conservation equation suffices to describe their effects. With a variety of influences acting on the charged particles, however, the effects may differ, depending on charge, size, and other physical and chemical attributes of the particle. Thus if several types (or species) of charged particles occur in an application, a separate conservation equation must be formulated for each species. This leads to a rich source of useful (or possibly annoying) interactions among the particles and external fields.

Generation of charged particles is essential, of course, but in practical work it is the generation rate, not the mechanism, which is important. If the rate is known, the behavior of a device like a radiation counter can be determined directly. Recombination receives a similar treatment in which the recombination coefficients are considered known and solution of the conservation equation gives the electrical response of the application, such as the ionospheric response to a solar flare.

Although generation and recombination are implicit in all problems, convection and diffusion are important only when the particles are distributed nonuniformly. Convection, or charge motion, carries patches of various density to other regions, so it has the effect of changing the number of charges there. This is put to practical use in discharging static electricity. Diffusion is a thermal effect which tends to spread the particles uniformly over the available volume. When the particles are charged, diffusion will be opposed by electric forces, which may lead to a nonuniform equilibrium distribution, as in the p-n junction.

6.1 PARTICLE CONSERVATION
(Ionization Gauge)

Summary

In many applications the behavior of individual charged particles holds little interest. Instead, the collective result of the motion of many charges is the key result. This type of application requires a method of keeping track of groups of particles in terms of densities and fluxes, as in an ionization gauge.

Theory

Consider a region of space containing many individual moving particles, as shown in Figure 6.1.1. At any instant the number of particles inside the volume is changing for several reasons. Some particles are crossing the sides, either entering or leaving. Since we are concerned only with a particular species of particles (electrons, for example), strict conservation does not apply here, and new particles may be generated inside the volume (by ionization). Likewise, particles may recombine inside the volume, causing a decrease in the number. In words, particle conservation can be stated as

$$\text{net increase} = \text{generation} - \text{recombination} - \text{loss through sides} \qquad (6.1.1)$$

The net increase is usually written in terms of the number density (n particles/m^3) integrated over the volume,

$$\text{net increase} = \frac{\partial}{\partial t} \iiint n \, d\mathcal{V} \qquad (6.1.2)$$

In writing these conservation equations, it is usually simpler to keep the volume

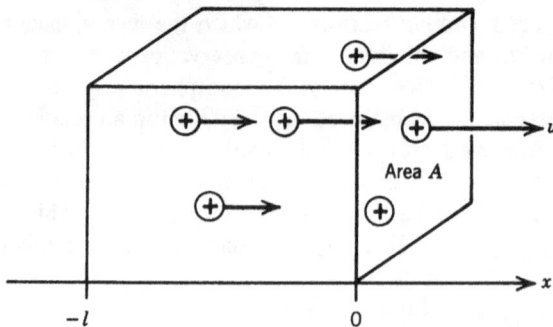

FIGURE 6.1.1. A volume with many moving particles.

(and its enclosing surface) fixed in space. With this restriction, the derivative can be brought inside the volume integral, giving

$$\text{net increase} = \iiint \frac{\partial n}{\partial t}\, d\mathcal{V} \tag{6.1.3}$$

The generation and recombination can be handled in a similar fashion, using the volume generation rate (G particles/m^3-s) and the volume recombination rate (R particles/m^3-s).

The last term in Eq. (6.1.1) is the net loss through the enclosing surface. This can be written as the integral of the outward particle flow over the enclosing surface as

$$\text{loss through sides} = \oiint \boldsymbol{\Gamma} \cdot d\mathbf{A} \tag{6.1.4}$$

The quantity $\boldsymbol{\Gamma}$ is called the particle flux vector. It points in the direction of particle movement, and its magnitude is given by the number of particles passing a unit area in a second. As an example of the flux vector, consider a parallelepiped with a length l and an area A. It is filled with particles with a density n, all of which are moving to the right with a velocity u. In a time $t = l/u$ all of the particles will pass through the end area, a total of nAl particles. The flux of particles is then

$$\boldsymbol{\Gamma} = \frac{nAl}{(l/u)A} = n\mathbf{u} \tag{6.1.5}$$

If the particles each have a charge q, the current density is given by

$$\mathbf{J} = q\boldsymbol{\Gamma} \tag{6.1.6}$$

The conservation equation for an individual species, Eq. (6.1.1), can now be written in terms of these integrals as

$$\iiint \frac{\partial n}{\partial t}\, d\mathcal{V} = \iiint G\, d\mathcal{V} - \iiint R\, d\mathcal{V} - \oiint \boldsymbol{\Gamma} \cdot d\mathbf{A} \tag{6.1.7}$$

when the surface is fixed.

The flux integral is a surface integral, whereas all of the other integrals in the particle conservation equation are volume integrals. We can convert the area integral by using Gauss' theorem as

$$\oiint \boldsymbol{\Gamma} \cdot d\mathbf{A} = \iiint (\nabla \cdot \boldsymbol{\Gamma})\, d\mathcal{V} \tag{6.1.8}$$

Since all of the integrals are now taken over the same fixed volume, the particle conservation equations may be written in terms of the integrands alone as

$$\frac{\partial n}{\partial t} + \nabla \cdot \boldsymbol{\Gamma} = G - R \tag{6.1.9}$$

This is the most common form of the equation, although we also use the integral form in some simpler applications.

Application: Air Ion Gauges

Much of the work in physical meteorology dwells on ions in the air and their effect on processes like cloud and rain formation. This work requires measurement of the ion generation in the atmosphere, often by collecting all of the ions formed in a known volume by means of a strong electric field. A schematic diagram of such an ion gauge is shown in Figure 6.1.2.

Ions are created, usually as pairs of opposite sign, in response to thermal excitation or ionizing radiation inside the volume. The electric field set up between the electrodes pulls the ions toward the electrodes, producing an electric current. To use such a gauge, we have to know the relation between the ion generation rate and the output current.

Application of Theory

For simplicity we calculate only the current flow of positive ions, which will flow to the right under the influence of the strong electric field. This field is usually strong enough in the gauge that every ion is immediately swept to the wall as soon as it is formed, so that the ion density stays close to zero all the time and, in any case, is constant. Under these steady-state conditions

$$\frac{\partial}{\partial t} \iiint n \, d\mathcal{V} = 0$$

Because of the quick removal, there is no time for the ions to recombine, so

$$R = 0$$

Thus the integral form of particle conservation, Eq. (6.1.7) may be written

$$0 = \iiint G \, d\mathcal{V} - \iint \boldsymbol{\Gamma} \cdot d\mathbf{A}$$

with uniform generation ($G = G_0$) over the volume the first term is

$$\iiint G_0 \, d\mathcal{V} = G_0 Al$$

The flux of positive ions is directed to the right wall, so

$$\oiint \boldsymbol{\Gamma} \cdot d\mathbf{A} = \Gamma A$$

and the equation becomes

$$G_0 Al = \Gamma A \qquad \text{or} \qquad \Gamma = G_0 l$$

Since the ions are charged,

$$J = q\Gamma = qG_0 l$$

and the total current to the right electrode is

$$i = JA = q(lA)G_0$$

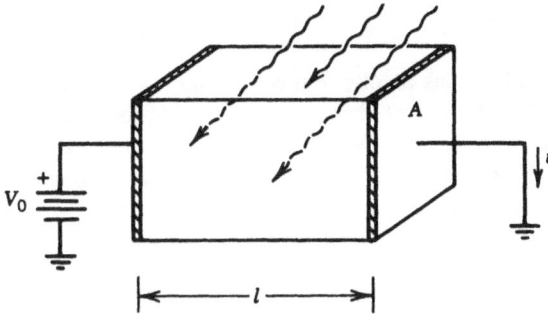

FIGURE 6.1.2. Air ionization gauge.

This current is proportional to the volume generation rate, and to known constants, so the gauge gives a linear relation between output current and generation. Typical parameters for the atmosphere are $G_0 \simeq 10^7$ ion pairs/m^3-s while the gauge might have a volume of $Al \simeq 10^{-5}$ m^3. For singly charged ions the current output would be

$$i = q(Al)G \simeq 0.016 \text{ fA}$$

Discussion

The current just calculated is only part of the answer since there will be at least one other charged species of opposite sign formed by the generation process. All of the other ions will supply current to the output, and the contributions of individual ions can not be separately measured with this device alone.

An interesting result of the analysis is the insensitivity of the output to the size of the electric field or to the speed of the ions. As long as the ions are removed before they can recombine, the output current remains steady regardless of the applied voltage. This effect, known as saturation, is fairly common in electrostatic devices. When it occurs, Ohm's law is meaningless, and any intuition gained in working with linear circuits must be disregarded lest false assumptions be made.

6.2 TRANSIENT CHARGE DECAY
(Decay of Ionospheric Disturbances)

Summary

Once charge carriers have been created by ionization, they do not remain indefinitely available. If positive and negative charges happen to come together, they may recombine, thus withdrawing two carriers from circulation and reducing the ability of the medium to conduct electricity. This is illustrated by the decrease in carrier density in the ionosphere after the end of a solar flare.

Theory

All electrostatic applications involving recombination are complicated by the need to keep track of several kinds of charge carriers during the course of events. In the simplest case there is only a single negative carrier with a density n_n and a single positive carrier with a density n_p. If carriers are not being generated in the bulk, conservation equations for each species may be written as

$$\frac{dn_n}{dt} = -\beta n_p n_n \tag{6.2.1}$$

$$\frac{dn_p}{dt} = -\beta n_p n_n \tag{6.2.2}$$

assuming two-particle recombination. The parameter β is the recombination coefficient. Many other recombination expressions are found in practice, depending on the details of the mechanism.

Both of these equations have the same recombination terms, since the carriers are lost in pairs. This does not mean, however, that there must be equal numbers of positive and negative carriers. If the positive carriers are initially in excess, so that there is a net positive charge density, the excess will remain, even though the number of carriers is decreasing. On the other hand, if the carrier densities are initially equal, then they will always be equal if there are no external forces such as electric fields or gravity to separate the charges from each other. By subtracting Eq. (6.2.1) from Eq. (6.2.2) we obtain the formal statement of these results as

$$\frac{d}{dt}(n_p - n_n) = 0 \tag{6.2.3}$$

or

$$n_p = n_n + \text{const} \tag{6.2.4}$$

If we assume a single source of ionization, so that $n_n = n_p = n$, the decay of ionization can be predicted from a solution of the equation

$$\frac{dn}{dt} = -\beta n^2 \tag{6.2.5}$$

with the initial condition

$$n(t = 0) = n_0 \tag{6.2.6}$$

This equation is similar to the equation for unipolar charge decay in the mobility limit. Its solution is

$$n = \frac{n_0}{1 + \beta n_0 t} \tag{6.2.7}$$

which is sketched in Figure 6.2.1. This is not exponential decay, and the decay time $t_e = (\beta n_0)^{-1}$ depends on the density initially present. Thus the ion content decays quickly when it is initially high, but it decays slowly when the ion content is low.

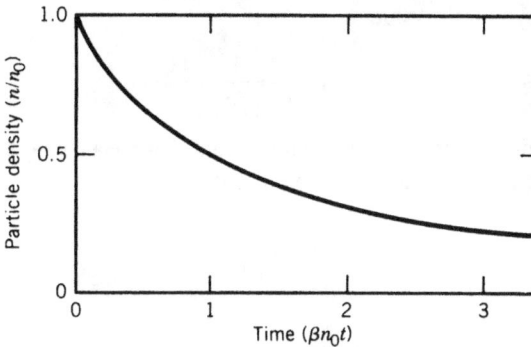

FIGURE 6.2.1. Decay by bimolecular recombination.

Example: Solar Flare Aftermath in the Ionosphere

As an example of simple recombination, consider the aftermath of a solar flare in the ionosphere. When such flares occur, they emit large amounts of ionizing radiation, some of which strikes the upper atmosphere of the earth, causing large increases in the ion density of the ionosphere. This ionization can interfere with radio communication and has also been blamed for disruptions of the electric power grid which can lead to blackouts. These effects persist for some time after the flare stops, since the decay of ionization by recombination proceeds relatively slowly. This time is therefore of interest in radio communications.

Application of Theory

The density of ions or electrons in the ionosphere at the end of a solar flare might typically be

$$n_0 \simeq 10^9/m^3 \tag{6.2.8}$$

while the recombination coefficient is

$$\beta \simeq 2 \times 10^{-13} \ m^3 s^{-1} \tag{6.2.9}$$

so that the electron density (and radio disturbances) will decrease substantially after a time

$$\tau \simeq 5000 \ s \ (\sim 1\tfrac{1}{2} \ hr) \tag{6.2.10}$$

Discussion

The initial decay time, or recombination time,

$$t_e = \frac{1}{\beta n_0} \tag{6.2.11}$$

depends on the density of carriers which is left after the flare; it is shortest for large densities. Long after the source of ionization is removed, the charge density approaches the asymptotic value

$$n \longrightarrow \frac{1}{\beta t} \qquad (6.2.12)$$

which is independent of the initial density. Thus if we wait long enough, we can guarantee that the carrier density will be less than

$$n < \frac{1}{\beta t} \qquad (6.2.13)$$

6.3   CHARGE CONVECTION AND CHARACTERISTICS
(Static Electricity Neutralizers)

Summary

When charged particles move, they can bring changes in density which are not caused by generation or recombination, but by convection. The basic solution method for charge convection problems is the theory of characteristics, illustrated here by a static neutralizer which involves the generation of charge in a nozzle and delivery of the charge by a high velocity air stream.

Theory

The examples of charge flows covered so far have been limited to devices in which the charges were uniformly distributed, so that their motion did not lead to changes in density. In this section the approach to problems involving nonuniform velocities and charge distributions is developed. It is based on the differential form of the conservation equation,

$$\frac{\partial n}{\partial t} + \nabla \cdot (n\mathbf{u}) = G - R \qquad (6.3.1)$$

for a single charge species. If more types of charge are present, a new equation must be written for each new species, but since all of the equations have the same form, the approach to the solution is similar.

    It is usually more convenient to make use of the vector identity

$$\nabla \cdot (n\mathbf{u}) = n(\nabla \cdot \mathbf{u}) + (\mathbf{u} \cdot \nabla)n \qquad (6.3.2)$$

to rewrite the equation in the form

$$\frac{\partial n}{\partial t} + \mathbf{u}(t, \mathbf{x}) \cdot \nabla n = G - R - (\nabla \cdot \mathbf{u})n \qquad (6.3.3)$$

Thus changes in the charge density, represented by the terms on the left, are caused by generation, recombination, and flow. Because derivatives of both time and space appear on the left, this is a partial differential equation for the charge density. Although the equation is not difficult, it does require techniques mark-edly different from those needed for the Laplace equation which usually arises in electrostatics.

The basic step in the solution of this equation is to recognize that the left-hand side of the equation has the form of a chain rule derivative. If the charge density is a function of a new variable s, which in turn depends on both time and space, the derivative of density can be written as

$$\frac{dn}{ds} = \frac{\partial n}{\partial t}\frac{dt}{ds} + \frac{\partial n}{\partial x}\frac{dx}{ds} + \frac{\partial n}{\partial y}\frac{dy}{ds} + \frac{dn}{dz}\frac{dz}{ds} \tag{6.3.4}$$

The left side of the conservation equation, Eq. (6.3.3), can therefore be written as

$$\frac{\partial n}{\partial t} + (\mathbf{u} \cdot \nabla)n = \frac{dn}{ds} \tag{6.3.5}$$

if the coefficients are comparable. This condition gives the relation between the new variables and the old variables in time and space as

$$\frac{dt}{ds} = 1 \tag{6.3.6}$$

$$\frac{dx}{ds} = u_x, \qquad \frac{dy}{ds} = u_y, \qquad \frac{dz}{ds} = u_z$$

which are called the characteristic equations for the problem. If these relations hold, the conservation equation can be written in the simpler form

$$\frac{dn}{ds} = G - R - (\nabla \cdot u)n \tag{6.3.7}$$

This is no longer a partial differential equation for the density but a first-order or-dinary differential equation, which is much simpler to solve. Its solution is given in terms of s, of course, and must be converted back into time and space variables by using the characteristic equations.

The characteristic equations also give a great deal of physical insight into the behavior of the solution and are often the primary topic of discussion when solv-ing charge convection problems. The relation between s and time is particularly simple and can be integrated to give

$$s = t - t_0 \tag{6.3.8}$$

The remaining equations, which give position in terms of s, have the form

$$x - x_0 = \int_{s=0}^{s} u_x(s)\,ds \tag{6.3.9}$$

Eliminating s from these equations gives a relation between time and space which is the trajectory that the system follows as s increases. Some typical paths are shown in Figure 6.3.1.

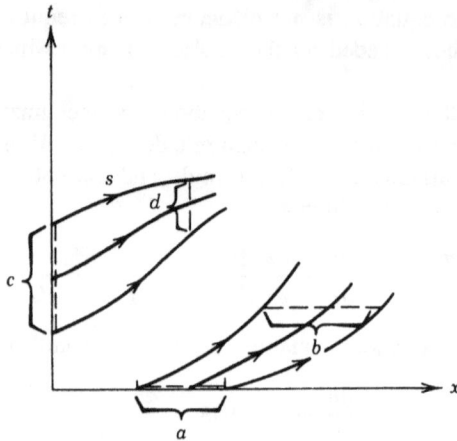

FIGURE 6.3.1. Characteristic paths.

Since the solution by characteristics is not usually covered in undergraduate curricula, it is worthwhile to outline the procedure to be followed in using these results to find the charge density. Because time and s are proportional, all of the systems which satisfy these equations are timelike, and the only conditions which can be imposed on the solution are initial conditions at the lowest value of s. For example, in Figure 6.3.1 the charge density could be specified at $t = 0$ along the line a, which covers a finite length of the system. The charge distribution at later times is then obtained by following each characteristic in the direction of increasing time s (upward), while solving the charge convection equation, Eq. (6.3.7). The characteristics which originate on either side of the segment a start with zero charge, and, in the absence of some later source of charge, will remain with zero charge. Only the characteristics which originate along the line a carry a charge, so the position of the packet at later times is obvious, once the characteristics have been drawn. For example, the charge packet will move and spread to cover the segment b at some later time.

A second type of initial condition common in charge convection gives the charge density over a period of time at some upstream point in the charge flow, like the segment c in Figure 6.3.1. Again, the later (downstream) behavior of the charge packet follows from the characteristic curves. In this example the initial pulse lasts for some time, but, as it travels downstream, the two characteristics which originate from the beginning and end of the pulse converge, which indicates that the pulse is compressed. Again, the actual value of the charge density at the downstream point is obtained by solving the charge conservation equation in terms of the characteristic variable s.

In practice, characteristics problems tend to break into two categories: easy and hard. The easy ones involve characteristic curves which can be obtained in advance of the charge density solution. They can therefore be determined once and for all at the beginning of the problem, which then reduces to a first-order integral for the charge density along the lines. The hard problems involve characteristic lines which are affected by the charge density. If the charge density changes, either at the initial point or during the response, the lines will move around and must therefore be recalculated continuously as the solution proceeds. In electrostatics these types of problems are particularly difficult since the charge trajectories are affected directly by the electric fields, which are in turn related to the charge density by Poisson's equation. Thus, at any instant of time, Poisson's equation must be re-solved for the entire region under consideration. The time and effort involved usually limit the solution to a single space dimension, even when large computers are available.

Example: Ion Generators for Static Discharge

Static electricity has always been a problem in industries which involve insulators moving past stationary objects. Typical examples include photographic film, printing presses, and oil pumping. Lately, the widespread use of highly insulating materials has exacerbated the problem, with more modern examples including the gates of field effect transistor (FET) semiconductor devices and plastic (principally polyester and nylon) clothing. These static buildups are always annoying and may damage or destroy the product, so much effort has been devoted to removing the charge as quickly as possible.

One common device for neutralizing the charge over a work station consists of a needle at high voltage in an air nozzle. As ions form at the surface of the needle, they are swept away by the air flow out the nozzle, where they can be directed toward the charged surface by the air flow. For the greatest effect, the initial charge density from the nozzle should be as high as possible, and efficient neutralizers should be designed to minimize loss of charge inside the nozzle.

Application of the Theory

The charge at the output of the nozzle can be calculated from the charge convection equations given previously. For simplicity, the nozzle is modeled as a tube of uniform radius a and length l, as shown in Figure 6.3.2. The air velocity is uniform across the tube, with a velocity U_0 directed entirely in the z direction. If the charge is unipolar, only one species need be considered. Its conservation equation in the steady state is

$$u_r \frac{\partial n}{\partial r} + u_z \frac{\partial n}{\partial z} = -(\nabla \cdot \mathbf{u})n \qquad (6.3.10)$$

where the charge velocity

$$\mathbf{u} = U_0 \mathbf{i}_z + \mu \mathbf{E} \qquad (6.3.11)$$

consists of a convection velocity as it is carried along by the air, in addition to a motion through the air due to electric fields.

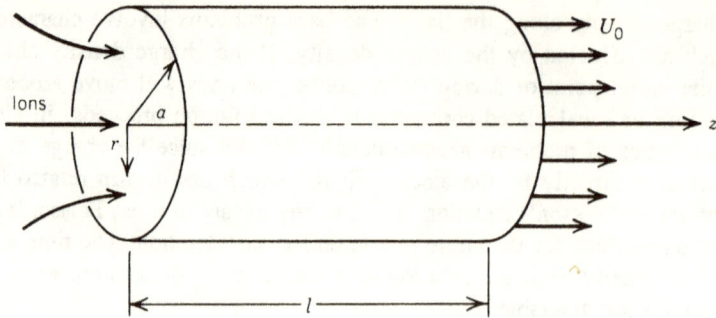

FIGURE 6.3.2. Ion source nozzle.

The left side of the equation has the standard form, while the right side contains divergence terms arising from the air flow and the electric field

$$\nabla \cdot \mathbf{u} = \nabla \cdot (U_0 \mathbf{i}_z) + \nabla \cdot (\mu \mathbf{E}) \tag{6.3.12}$$

The air velocity is uniform, so its divergence, which is a space derivative, vanishes, leaving only the divergence of the electric field. From Gauss' law, this is

$$\nabla \cdot (\mu E) = \mu (\nabla \cdot E) = \frac{\mu n q}{\epsilon} \tag{6.3.13}$$

if the mobility is uniform. Charge convection, Eq. (6.3.10), can now be written as

$$\frac{dn}{ds} = -\frac{\mu q n^2}{\epsilon} \tag{6.3.14}$$

by using the characteristic formulation. This equation has the solution

$$n(s) = \frac{n_0}{(1 + \mu n_0 q s / \epsilon)} \tag{6.3.15}$$

in terms of the initial particle density n_0. Note that we have obtained the charge density solution without even mentioning the characteristic curves. The charge always decreases along the characteristic curve (increasing s), so the highest outputs will be obtained when the least time elapses between creating the charge and using it.

The solution is worthless, however, unless the characteristic variable s can be converted back into real time and space. The characteristic curves, which show where the charge is going, are obtained by solving

$$\frac{dz}{ds} = U_0 + \mu E_z \tag{6.3.16a}$$

$$\frac{dr}{ds} = \mu E_r \tag{6.3.16b}$$

Since the primary purpose of the device is to blow the charge out of the nozzle, we can assume that the velocity generated by the electric field is much less than the air velocity and neglect the last term in Eq. (6.3.16a). This gives the relation between z and s as

$$z = U_0 s \tag{6.3.17}$$

and the charge density in terms of distance from the inlet as

$$\rho = \rho_0 \left[1 + \frac{\mu \rho_0 z}{\epsilon U_0} \right]^{-1} \tag{6.3.18}$$

Plots of the density decay are shown in Figure 6.3.3. The decay shown here is not the common exponential decay and has some peculiar features. One factor of extreme importance occurs downstream, when all of the curves come together regardless of how much initial charge was injected. If the distance is sufficiently long ($z \gg \epsilon U_0/\mu\rho_0$), the density reduces to $n \rightarrow \epsilon/\mu t$, which is independent of the initial charge. If a given time has elapsed, the charge can never exceed the limiting expression, regardless of how much was initially injected. The excess charge merely creates stronger space-charge fields, which drive it to the wall faster. Thus faster transport, not stronger injection, is the key to improving this type of charge generator.

The radial electric field in Eq. (6.3.16b) can be obtained approximately by assuming that the nozzle is so thin that all the field lines from the space charge go directly to the walls rather than along the nozzle. This makes the electrostatic field one dimensional, so that straightforward solution of Poisson's equation in a cylindrical geometry gives

$$E_r \simeq \frac{\rho(s)r}{2\epsilon} \tag{6.3.19}$$

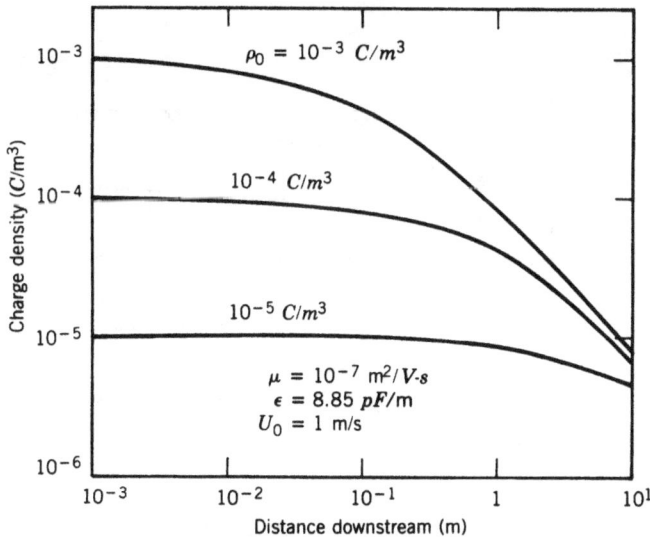

FIGURE 6.3.3.   Charge decay in convection.

Substituting this field into the characteristic equations gives

$$\frac{dr}{ds} = \frac{\mu\rho_0 r}{2\epsilon}\left(1 + \frac{\mu\rho_0 s}{\epsilon}\right)^{-1} \tag{6.3.20}$$

which can be integrated in the form

$$\int_{r_0}^{r} \frac{dr}{r} = \frac{\mu\rho_0}{2\epsilon}\int_{s=0}^{s=z/U_0}\left(1 + \frac{\mu\rho_0 s}{\epsilon}\right)^{-1} ds \tag{6.3.21}$$

The limits correspond to a trajectory which starts with a radial position r_0 at the inlet $(s = 0)$. Carrying out this integral gives the trajectory as

$$r = r_0\sqrt{1 + \frac{\mu\rho_0 z}{\epsilon U_0}} \tag{6.3.22}$$

Several of these trajectories are sketched in Figure 6.3.4.

These paths clarify the mechanism by which the charge decreases, since they show how the charge packet tries to expand under the influence of electrostatic forces as it is swept down the nozzle. Where its radial velocity carries it to the wall, it is lost from the flow. Note that this effect is strongest near the entrance, where the net charge is largest.

Discussion

The characteristic equations arise whenever charge flows. This happens when they are blown about, as in the static neutralizer, but it also happens whenever they are placed in an electric field. Since electric fields are ubiquitous in electrostatics, the techniques presented here are quite commonly used. They are especially helpful when several species of charge are present, such as the holes and electrons in

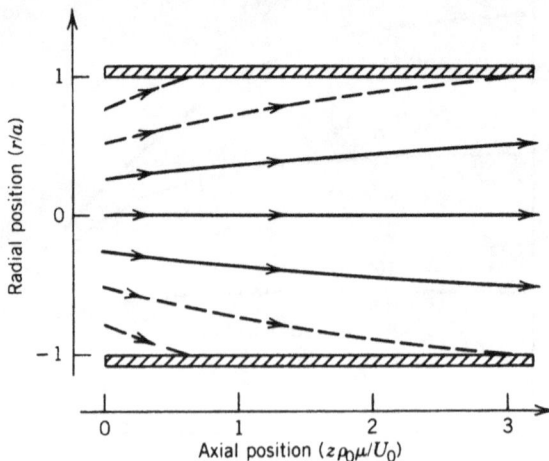

FIGURE 6.3.4. Charge trajectories under the influence of space charge: ———, charges that leave the nozzle; – – –, charges that strike the wall.

semiconductor devices or the positive and negative ions in gaseous and liquid insulators. They are well suited to numerical techniques because their timelike nature gives the results as the solution of ordinary differential equations with initial conditions, which can usually be handled by a straightforward stepping technique requiring little memory.

6.4 DIFFUSION FLUX
(Semiconductor Junction Potential)

Summary

Particle fluxes are driven by thermal kinetic effects, in addition to external forces such as electric fields or convection. This thermal motion gives rise to the diffusion of the individual species in directions determined by their density gradient. Diffusion plays a key role near interfaces, such as the p-n junction, which is described in the example that follows.

Theory

Before, particle fluxes were implicitly assumed to result only from externally directed motions of particles caused by electric fields, convection, and other external force fields. A net flux can also occur if the particle density is nonuniform and kinetic processes important. Consider the net flux crossing the area located at x in Figure 6.4.1 in the absence of external fields. To simplify the discussion, we restrict the motion to a single direction. Each particle moves with a speed u_t, which in a more complete analysis would be related to temperature. Each particle collides every t_c seconds and is equally likely to be moving in the positive or negative x direction after the collision. The figure shows the situation immediately after all

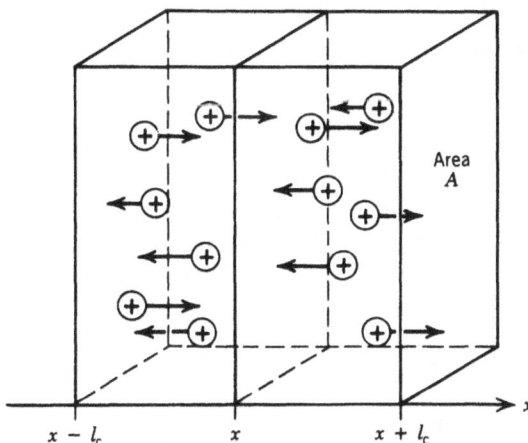

FIGURE 6.4.1. Model for diffusion flux.

of the particles have just collided. To the left of the area, one-half are traveling to the right after the collision and will cross the surface if they are closer than one collision length

$$l_c = u_t t_c \tag{6.4.1}$$

Thus one-half of the particles in the volume Al_c will cross in t_c seconds, giving a flux contribution to the right

$$\frac{1}{2} n\left(x - \frac{l_c}{2}\right) \frac{l_c A}{(t_c A)} \tag{6.4.2}$$

where the particle density is evaluated at the center of the region which supplied the flux. The ratio $l_c/t_c = u_t$ is the thermal velocity of the particles, which depends mostly on the temperature and particle mass. Thus the flux to the right can be rewritten as

$$\Gamma_r = \frac{1}{2} n\left(x - \frac{l_c}{2}\right) u_t \tag{6.4.3}$$

Likewise, a left-directed flux of

$$\Gamma_l = \frac{1}{2} n\left(x + \frac{l_c}{2}\right) u_t \tag{6.4.4}$$

can be defined.

Combining these two components gives the net flux as

$$\Gamma = \Gamma_r - \Gamma_l = \frac{1}{2} u_t \left[n\left(x - \frac{l_c}{2}\right) - n\left(x + \frac{l_c}{2}\right) \right] \tag{6.4.5}$$

With relatively short collision lengths, the density on the right can be approximated by Taylor series as

$$n\left(x + \frac{l_c}{2}\right) \simeq n(x) + \frac{\partial n}{\partial x} \frac{l_c}{2} \tag{6.4.6}$$

with a similar expression for the left. Substituting into the net flux Eq. (6.4.5) yields

$$\Gamma = \frac{-l_c u_t}{4} \frac{\partial n}{\partial x} \equiv -D \frac{\partial n}{\partial x} \tag{6.4.7}$$

The diffusion coefficient D is related to the thermal kinetic properties of the particle. It can often be estimated from the Einstein relation $D \approx \mu kT/q$. The diffusion flux should be added to any other particle fluxes which exist to find the total flow under a given set of circumstances.

Example: Equilibrium in a p-n Junction

Most of the solid-state devices used today depend to some extent on the electrostatic properties of the interface between two materials which are almost identical except for the number and type of free charge carriers. The two sides are usually

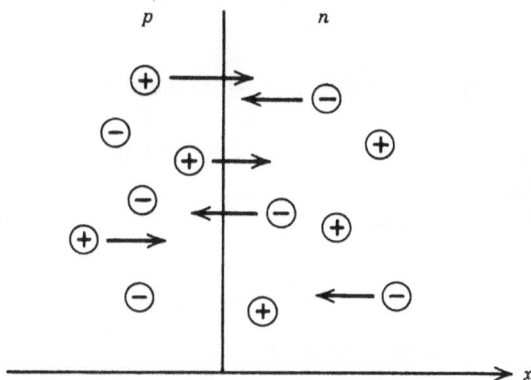

FIGURE 6.4.2.   A p-n junction.

the same basic chemical compound (e.g., silicon), but they have been treated with different impurities so that in one the mobile carriers are mostly electrons (n type) whereas in the other the carriers are mostly "holes," or vacancies left by a deficit of electrons (p type). The interface between these two regions is called a p-n junction.

Since the carriers are mobile, they can move back and forth across the junction. The positive carriers (holes) can cross from the p region, as shown in Figure 6.4.2, into the n region, where they form a net positive charge. Since the holes are densely concentrated in the p region, there will be a net diffusion toward the n region where they are scarce. As the positive charges build up in the n region, their net space charge will repel any additional holes diffusing in that direction, and, if the process continues, a balance between the diffusion and the repulsion of the electrostatic charges will develop. Since this is an equilibrium, each charge carrier must separately satisfy this balance condition.

The equilibrium depends on the number of carriers in each region and also on the voltage difference across the junction, which is related to the repulsive field opposing the diffusion. A knowledge of this relation is essential to understanding and designing virtually all solid-state devices, including transistors, FETs, silicon-controlled rectifier (SCRs), and the like.

Application of the Theory

The net flux of each species across the junction is driven by diffusion and by the electric field. For the positive species (holes) the flux is

$$\Gamma = -\mathbf{D}\frac{\partial n}{\partial x} + \mu En \tag{6.4.8}$$

Since the junction is in equilibrium, the conservation equation for the holes reduces to the steady-state form

$$\nabla \cdot \Gamma = \frac{d}{dx}\left(-D\frac{dn}{dx} + \mu En\right) = 0 \tag{6.4.9}$$

which can immediately be integrated to

$$\Gamma = -D\frac{dn}{dx} + \mu En = \text{constant} = 0 \tag{6.4.10}$$

The constant of integration is zero because in equilibrium there is no net flow of any individual species.

The electric field in the junction is related to electric potential by

$$E = -\frac{d\Phi}{dx} \tag{6.4.11}$$

so the flux balance equation can be rewritten as

$$D\frac{dn}{dx} = -\mu n\frac{d\Phi}{dx} \tag{6.4.12}$$

which can be integrated across the whole region of the junction to give

$$\int_{x=-\infty}^{\infty}\frac{dn}{n} = -\frac{\mu}{D}\int_{\Phi=0}^{V} d\Phi \tag{6.4.13}$$

or

$$n(\infty) = n(-\infty)e^{-\mu V/D} \tag{6.4.14}$$

Thus there is a simple relation between the voltage drop across the junction and the relative number of holes on either side of it.

Discussion

Strictly speaking, the relation between voltage and carrier densities only holds when the junction is in equilibrium, carrying no current. This condition is of little interest in most applications of semiconductor devices, but it turns out that the relation just derived is remarkably accurate even when the junction is not in equilibrium. This occurs because the net flux in Eq. (6.4.10) is the difference between two relatively large partial fluxes. For example, if the p region has a hole density of $10^{18}/\text{cm}^3$ and a diffusion constant of 50 cm^2/s, the current caused by diffusion alone will be approximately

$$J_{\text{diff}} = -q\frac{\Delta n}{\Delta x} = 800,000 \text{ A/cm}^2$$

across a junction which is 0.1 μm wide. This is an enormous value, but it is balanced by an equally large flow of charge in the reverse direction caused by the repulsive electric field. If the junction carries a net current, one of these partial currents must be slightly greater, but for the currents typically carried by junctions ($\sim$1 A) the unbalance is too slight to seriously change the equilibrium equations. Thus the junction remains in approximate equilibrium under all reasonable current levels in ordinary junctions. Of course, if the hole concentration were reduced, or the junction became very wide, the diffusion flux would be reduced, and less current would be needed to unbalance the junction.

Diffusion is important if the group $D/\mu V$ is large, so it tends to arise in regions of small voltage drop, such as inside conducting materials or at narrow boundary regions. In most materials, the ratio

$$\frac{D}{\mu} \simeq \frac{kT}{q_e}(\simeq 25 \text{ mV at } 20°C)$$

where q_e is the electronic charge. This is called the thermal voltage and is a useful yardstick for deciding whether diffusion is likely to play a role in a particular application.

BIBLIOGRAPHY

Aris, R., and N. R. Amundson, *Mathematical Methods in Chemical Engineering*, Vol. 2, Prentice-Hall, Englewood Cliffs, NJ, 1973.

Bockris, J. O'M., and A. K. N. Reddy, *Modern Electrochemistry*, Plenum, New York, Vol. 2, 1970, Chap. 7.

Chalmers, J. A., *Atmospheric Electricity*, Pergamon, Oxford, 1967.

Croitorou, Z., Space charges in dielectrics, *Prog. Dielectr.*, **6**: 103–146.

Gray, P. E., and C. L. Searle, *Electronic Principles*, Wiley, New York, 1969, Chap. 4.

Hughes, A. L., and L. A. DuBridge, *Photoelectric Phenomena*, McGraw-Hill, New York, 1932, Chap. 8.

Israel, H., *Atmospheric Electricity*, Vol. 1, Israel Program for Scientific Translations, Jerusalem, 1971.

Lion, K. S., *Instrumentation in Scientific Research*, McGraw-Hill, New York, 1959.

Melcher, J. R., Electric fields and moving media, *IEEE Trans. Educ.*, **E-17**: 100–110 (1974).

Rose, D. J., and M. Clark, *Plasmas and Controlled Fusion*, Wiley, 1961, Chap. 8.

Schokley, W., *Electrons and Holes in Semiconductors*, Van Nostrand, New York, 1950, Chap. 12.

Simon, F. N., and G. D. Rork, Ionization-type smoke detectors, *Rev. Sci. Inst.*, **47**: 74–80 (1976).

Streetman, B. G., *Solid State Electronic Devices*, Prentice-Hall, Englewood Cliffs, NJ, 1972, pp. 149–175.

Whitten, R. C., and I. G. Poppff, *Fundamentals of Aeronomy*, Wiley, New York, 1971, Chap. 10.

Wilkenson, D. H., *Ionization Chambers and Counters*, Cambridge University Press, Cambridge, 1950.

Woodson, H. H., and J. R. Melcher, *Electromechanical Dynamics*, Wiley, New York, 1968, Chap. 7.

PROBLEMS

PROBLEM 1 (COMMUNICATIONS)

A solar flare begins to generate ion pairs in the ionosphere at a constant rate G_0. If the initial ion density is n_0, find the ensuing density $n(t)$, when the recombination follows a binary law, $R = \alpha n_+ n_-$.

PROBLEM 2 (INSTRUMENTATION)

A solid-state photodetector consists of a parallelpiped of length l ($= 1$ cm) and area A ($= 1$ cm^2). A voltage V ($= 10$ V) is applied to electrodes at the two ends. The photodetector is uniformly illuminated by light which generates G ion pairs/ m^3-s. The ions move to the electrodes in the mobility limit with mobilities μ_+ and μ_-. As they drift, they are captured by trapping centers which remove them from the flow at a rate $R_+ = \beta_+ n_+$, $R_- = \beta_- n_-$. Space-charge fields can be neglected.

  a. Find the ion densities $n_+(x)$ and $n_-(x)$ in the steady state.

  b. If the light is turned off ($G = 0$) at $t = 0$, find the resulting transient densities $n_+(x,t)$ and $n_-(x,t)$ and sketch versus x at $t = l^2(2\mu V)^{-1}$.

PROBLEM 3 (SMOKE DETECTOR)

A smoke detector consists of a chamber filled with room air which is uniformly ionized by a radioactive source. This source generates G_0 ion pairs m^{-3} s^{-1}. A voltage V ($= 10$ V) is applied between two parallel electrodes separated by a distance d ($= 1$ cm) so as to collect these ions and measure the current. Assume that the number of ions generated is too small to affect the electric field. The ionic flux is driven by mobility and diffusion. Find the ionic fluxes at the electrodes. Both electrodes absorb ions readily, so the ion density there vanishes.

$$n(x = 0) = n(x = d) = 0$$

PROBLEM 4 (BIOMEDICAL)

When a protein is put in water, it often behaves as if it has a net charge and is surrounded by a blocking layer. If an electric field is applied, the material moves, giving rise to an effect called electrophoresis, which is routinely used in the medical analysis of blood. The apparent charge is caused by diffusion effects at a junction.

If the initial densities of positive and negative ions are equal throughout the water $n_+(\infty) = n_-(\infty) = n_0$ and a spherical partical, with a radius a and a potential V, is immersed in the water:

  a. Find the negative and positive ion densities $n_+(r)$, $n_-(r)$ in terms of the local potential $\phi(r)$ in the water.

  b. Show that Poisson's equation has the form

$$\frac{1}{r^2}\frac{d}{dr}\left(r^2\frac{d\phi}{dr}\right) = K^2 \sinh\left(\frac{q\phi}{kT}\right)$$

  c. Approximate $\sinh q\phi/kT \simeq q\phi/kT$ and find $\phi(r)$ near the particle ($r > a$). (The solution to Poisson's equation will have the form $e^{\pm Kr}/r$).

  d. What is the Debye length (the distance over which the particle disturbs the ion distribution)?

PROBLEM 5   (ELECTRIC POWER)

A parallel electrode thermionic generator is attached to an external circuit which supplies a voltage V (= 100 V) to the electrodes. One electrode is heated to emit electrons with a density n_0 next to the electrode. These charges move in response to diffusion and to the electric field set up by the external voltage (neglect space-charge effects). Find and plot the electrode current in the steady-state current as a function of the applied voltage.

III

ELECTROSTATICS
OF MATERIALS

CHAPTER

7

CONDUCTION
AND
BREAKDOWN

The electric current that flows in response to external influences takes many forms in different applications. Sometimes the current is linearly proportional to the electric field, giving an ohmic conductivity (or resistivity) which can describe the situation under all operating conditions. More often, however, the relation is more complicated, so that the current has a complex, or even indeterminate, relation to the electric field. One example is the macroscopic particle current, which accounts for most of the leakage in gas-insulated power stations.

Whatever form the conduction takes, it can be described in terms of charge carriers moving through the medium. The methods of the last few chapters are usually employed to determine the number and velocity of each type of carrier and thus the current flow. In some cases a single charge species controls the current flow, which greatly simplifies device analysis, as illustrated by a proportional counter.

In general more than one charge species must be considered to understand current flow in electrostatics. When all of the carriers may have different properties, such as charge or velocity, a statistical approach based on distributions is usually needed. Even if only two species interact, however, many new modes of conduction are possible. One of these, electrical breakdown by means of impact ionization, is described in Section 7.3.

7.1 WHEN IS CONDUCTION OHMIC?
(Compressed Gas Insulation)

Introduction

Ohmic conduction is the exception rather than the rule in electrostatics applications, yet it is often assumed implicitly in discussions. In this section some of the conditions for ohmic conduction are formulated to furnish guidelines for practice. All of these guidelines are violated by the particle current which accounts for most of the leakage in compressed gas insulators.

Theory

In the steady state the current flow is described in terms of the charge and flux of the individual carrier species as

$$\mathbf{J} = \sum_i q_i \mathbf{\Gamma}_i \qquad (7.1.1)$$

There are several driving forces for the flux, such as diffusion, which are usually neglected when speaking of conductivity, so they are neglected here. If present, of course, they add to the flux generated by the electric field, but for simplicity we assume that the flux is related to the drift velocity and density of carriers by

$$\mathbf{\Gamma}_i = n_i \mathbf{u}_i \qquad (7.1.2)$$

In the mobility limit

$$\mathbf{u}_i = \mu_i \mathbf{E} \qquad (7.1.3)$$

Combining all of these relations gives the current as

$$\mathbf{J} = \left(\sum_i n_i q_i \mu_i \right) \mathbf{E} \equiv \sigma \mathbf{E} \qquad (7.1.4)$$

The quantity in parentheses is commonly called the conductivity and is independent of the electric field in the ohmic limit. Thus one way to ensure ohmic behavior is to require that n, q, and μ are each independent of $\mathbf{E}$.

We have seen numerous examples so far of exceptions to this rule. The carrier density, for instance, is frequently affected by the $\mathbf{E}$ field. This occurred in the radiation monitor when the carriers were generated at a finite rate and increasing the $\mathbf{E}$ field removed them more quickly, thus decreasing n. The opposite effect often occurs at electrodes, where increasing the electric field pulls more carriers into the conduction region, leading to an increase of n there. For ohmic behavior, effects such as these should be small. This is really the same as saying that there are so many carriers available that those removed or created by the electric field represent a small fraction of the carrier population. Thus good conductors, such as metals and water which have large numbers of carriers, tend to be ohmic, whereas poor conductors like plastic and air have few carriers to begin with so that n is more easily influenced by external effects like electric fields.

The charge q_i of the individual carriers should also be constant to ensure ohmic behavior. This is usually the case with electrons and ions, but when dealing with larger particles exceptions abound. One example is the current carried by particles bouncing between electrodes. These particles are charged by induction when in contact with the electrode, so their charge is proportional to the applied electric field. This makes the conduction nonohmic.

The mobility can also vary with applied field, especially in solids which transport carriers by hopping between potential wells. At low fields the mobility is very low because the particles can not leave the wells, but mobility increases with the stronger fields which are able to pull electrons up out of the wells in addition to moving them along to the next well.

Example: Particle Conduction in Gas Insulation

When high voltages exist in the close quarters of compact electric power substations, electronegative gases such as sulfur hexafluoride (SF_6) are often used to insulate the conductors. Such gases have much higher breakdown fields than does air, especially when compressed, and are free from the deterioration at high temperatures which often limits the usefulness of oils and solid polymers.

In gaseous-insulated equipment particle conduction accounts for most of the current leakage and breakdown, so it is necessary to understand and control such conduction to achieve the full potential of this approach. Particle conduction arises mostly from the small fragments left behind from the mechanical construction of the equipment. These particles offer a good example of nonohmic conduction since they violate every requirement listed before.

Application of the Theory

For conductivity (or ohmic conduction) to be a plausible way of expressing the conduction of current through a medium, the current density must be linearly related to the electric field. Conductivity involves the number of carriers, their charge, and their velocity, all of which may be influenced by E. Although all of these influences may combine to give a linear dependence, the most common set of restrictions for ohmic current flow involves density and charge which are both independent of field along with a linear velocity dependence described by a constant mobility. Let us test these assumptions for particle current.

The first assumption requires that the number of each carrier participating in the conduction remain constant. Initially these particles lie on the bottom of the device, held there by gravity, as shown in Figure 7.1.1. When a field is applied, a particle which is in contact with the lower electrode becomes charged and attracted toward the upper electrode. It can not move at all, however, until the electrostatic force becomes large enough to overcome the weight. Thus the number

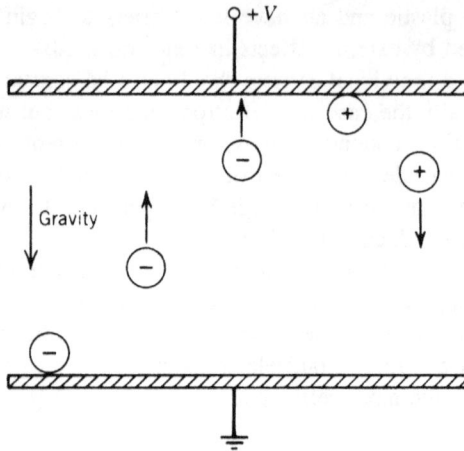

FIGURE 7.1.1. Particle current flow in compressed gas insulation.

density for a given particle type will be zero below a critical field and rise abruptly to include all particles of that type when the critical field is exceeded. This highly nonlinear dependence of number density $n(E)$ on applied field precludes an ohmic description of the current flow.

The second assumption requires the charge on the carrier to be constant. In particle conduction the charge arises by induction when the particle contacts the electrode. A stronger electric field will naturally induce more charge $q(E)$ so that the constant charge assumption is also violated. Once the particle leaves the electrode, it can still gain or lose charge by breakdown or collision with another particle or with the electrode. Even with a constant electric field, then, the charge on the particle is not constant. In some applications, averaging over all of the particles may be appropriate to reduce the variation caused by collisions and breakdown, but the electric field dependence still remains to violate the constant charge assumption.

A third assumption for ohmic current involves a charge velocity linearly related to electric field. As we have shown in Chapter 5, this requires that two conditions be met. The first is a short mechanical relaxation time, which allows a rapid acceleration to the drag-limited velocity. The relaxation time increases with particle mass and decreases with the drag of the medium. For macroscopic particles in a gas, therefore, the mechanical relaxation time is likely to be long, especially for the larger particles, and the average velocity of the carriers will be significantly influenced by acceleration effects.

For the smaller particles acceleration effects are less important, and the mobility expression $u = \mu E$ may be valid. The mobility, however, depends on the particle charge, with highly charged particles moving faster in the same field. This dependence of mobility $\mu(E)$ on the applied field further invalidates the ohmic assumptions. With all of these assumptions violated, it is clearly impossible to speak of a *conductivity* when discussing the leakage current due to particle motion.

Discussion

The example given here violates all of the guidelines for ohmic conduction. The situation is not usually this bad, but ohmic conduction is the exception, not the rule, in electrostatics applications. Each type of situation has its own characteristic limitations and sometimes a different name, such as the *electrode effect* in atmospheric science. Large carriers, for example, are likely to have variable charges due to collisions and breakdown during midflight, and their velocity is influenced more by inertial effects. Small particles are free of these limitations, but their mobility is influenced by the surrounding medium. This is especially noticeable in anisotropic media such as solid and liquid crystals.

All types of carriers are nonohmic when their number is small, since large fractional changes result from small external influences such as recombination or convection. For this reason the insulating materials which form an essential part of any electrostatic device are usually nonohmic, and handbook values of "conductivity" are often useless or misleading in practice, although they do indicate what may be achieved under certain well-controlled conditions.

7.2   CONDUCTION WITH A SINGLE SPECIES
(Radiation Counters)

Summary

In some conduction problems there is one charge species which controls the current flow, in the sense that all of the other flows can be determined from that one, but it is not influenced by the others. This greatly simplifies the model, as illustrated by the proportional radiation counter.

Theory

Generally there are several types of charge carriers which contribute to the total current flow, and these carriers often interact with each other, making a prediction of the total current quite complicated. In some cases, however, a single charge species dominates the dynamics so that knowledge of its motion leads directly to the behavior of the other species and hence to the total current. This can happen only if none of the terms in the conservation equation depends on any of the other charge species.

The conservation equation for each species n_i has the general form

$$\frac{\partial n_i}{\partial t} + \nabla \cdot \Gamma_i = G_i - R_i \tag{7.2.1}$$

The first term clearly depends only on the first species, so it never introduces coupling terms. The second term involves the flux, which is often given by

$$\mathbf{\Gamma}_i = -D_i\,\nabla n_i + n_i\mathbf{u}_i \tag{7.2.2}$$

Again, these terms depend primarily on n_i alone, so that coupling may usually be neglected. The drift velocity $\mathbf{u}_i$ may occasionally lead to complications since the velocity $u(E)$ is influenced by the electric field. Changes in charge carrier density will affect the field, which in turn affects the carrier density.

It is the generation G and recombination R, however, which are most likely to couple the conservation equations because each of these events involves at least three separate species. In recombination, for example, a positive carrier and a negative carrier combine to form an uncharged species. As a result the number of positive, negative, and neutral species will all change. In some situations, however, the effects of some of the changes on the number of a particular species is slight and may be safely neglected.

Example: Proportional Radiation Counter

In many areas of nuclear technology the detection of nuclear radiation plays a crucial role. One common detection method relies on the ionization left in the wake of an energetic particle as it passes between two electrodes. The ionization produced by a single particle is very small and difficult to detect, so it is usually amplified in the chamber itself by a process called avalanching.

An electron avalanche occurs when the applied field can accelerate an electron to very high velocities (or kinetic energies) between collisions with the background gas molecules. If the kinetic energy exceeds the ionization energy of the gas, the collision will create a new electron, along with a positive gas ion. The new and the old electron will be accelerated in the field, and the process will continue to repeat until the electrons reach the electrode. There will be many more electrons (and ions) than were created by the ionizing radiation, so the current pulse at the output will be much larger than it would have been without avalanching. The amplification due to the avalanche process is the primary design criterion for such counters.

Application of the Theory

The counter is modeled by a uniform region between two parallel electrodes as shown in Figure 7.2.1. The family history of a single photoionization event is shown schematically, but the total current will be due to numerous such events, which we assume to be occurring constantly. Three separate species influence the current flow in the counter, so three separate conservation equations must be written. For the electrons,

$$\frac{\partial n_-}{\partial t} + u_-\frac{\partial n_-}{\partial x} = G_0 + \beta_0 n_0 n_- - \beta_+ n_+ n_- \tag{7.2.3}$$

The generation consists of a constant term G_0 due to the radiation and an avalanche term $\beta_0 n_- n_0$ which depends on the number of electrons already present and on the

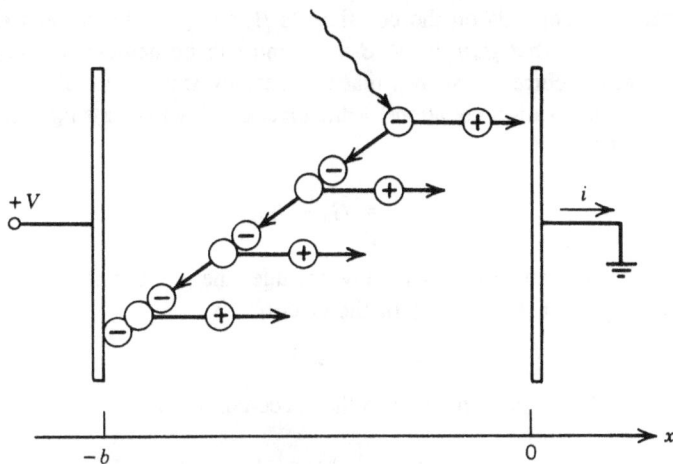

FIGURE 7.2.1. Avalanching in a proportional counter.

number of neutral molecules available for ionization. The positive ion density is given by

$$\frac{\partial n_+}{\partial t} + u_+ \frac{\partial n_+}{\partial x} = G_0 + \beta_0 n_0 n_- - \beta_+ n_+ n_- \tag{7.2.4}$$

which is identical to the electron equation except for the particle velocity u_+. Finally, the neutral molecule equation is

$$\frac{\partial n_0}{\partial t} + u_0 \frac{\partial n_0}{\partial x} = -G_0 - \beta_0 n_0 n_- + \beta_+ n_+ n_- \tag{7.2.5}$$

The generation and recombination terms have opposite signs because the creation of ions and electrons implies the loss of gas molecules.

Each of these three conservation equations involves all three particle densities, but some equations are more important than others. In most counters the density of neutral molecules is always much greater than the ion density, $n_0 \gg n_+$, $n_0 \gg n_-$; but the changes in all three densities are comparable, $|\delta n_0| = |\delta n_+| = |\delta n_-|$. Thus

$$\frac{\delta n_0}{n_0} \ll \frac{\delta n_-}{n_-} \quad \text{or} \quad \frac{\delta n_+}{n_+} \tag{7.2.6}$$

and the avalanche will have very little effect on the total number of gas molecules. With small fractional changes we can assume

$$n_0 \simeq \text{constant} \tag{7.2.7}$$

and ignore the conservation equation for the gas molecules.

Comparing the avalanche $\beta_0 n_0 n_-$ and recombination $\beta_+ n_+ n_-$ terms shows that recombination may be a relatively weak effect because $n_0 \gg n_+$. The relative size of these terms also depends on the coefficients β_0 and β_+, but in most radiation detectors it turns out that $\beta_+ n_+ n_- \ll \beta_0 n_0 n_-$ and can be neglected. Finally, we will assume that the current flow has reached a steady state. With all of these assumptions, the conservation equation for the electrons involves a single unknown, the electron density n_-.

$$u_- \frac{dn_-}{dx} = G_0 + \beta n_- \qquad (7.2.8)$$

The avalanche coefficient $\beta \; (= \beta_0 n_0)$ now includes the gas density.

In the mobility limit the velocity of the electrons is

$$u_- = -\mu_- E \qquad (7.2.9)$$

If the applied field is large compared to the space-charge field,

$$E = \frac{V}{b} = E_0 \qquad (7.2.10)$$

and the conservation equation reduces to

$$\frac{dn_-}{dx} = -\frac{G_0}{\mu E_0} - \frac{\beta}{\mu E_0} n_- \qquad (7.2.11)$$

which has a general solution of the form

$$n_- = \frac{-G_0}{\beta} + K \exp\!\left(\frac{-\beta x}{\mu E_0}\right) \qquad (7.2.12)$$

indicating that the carrier density increases exponentially across the gap in the direction of motion.

All of the electrons are swept to the left, but none are emitted from the right electrode. There can be no carriers just outside the right electrode since the volume in which they could have been generated by radiation is vanishingly small. Thus the appropriate boundary condition is

$$n_-(x = 0) = 0 \qquad (7.2.13)$$

which gives the solution for electron density as

$$n_- = \frac{G_0}{\beta}(e^{-x/l_c} - 1) \qquad (7.2.14)$$

The collision length in the exponential term,

$$l_c = \frac{\mu_- E}{\beta} \qquad (7.2.15)$$

depends on the electric field both explicitly and through the collisional ionization constant $\beta(E)$. It is often replaced by the single parameter called the first Townsend ionization coefficient.

The electron current which reaches the electrode in the steady state can be determined by evaluating the electron flux at the left electrode,

$$J_-(x = -b) = q\Gamma_-(-b) = |q|\mu_- E_0 n_-(x = -b) \qquad (7.2.16)$$

Since there are no ions reaching the left electrode, the total terminal current is

$$J = |q|Gl_c(e^{b/l_c} - 1) \qquad (7.2.17)$$

which is plotted in Figure 7.2.2 as a function of electrode spacing. From this figure, it is clear that the current increases as the separation between the electrodes is made larger, and that the current is always greater than the saturation value

$$J_{sat} = qG_0 b \qquad (7.2.18)$$

obtained when all the primary carriers are collected before they can recombine or cause avalanches.

Discussion

If this device is to be used to detect the ionization produced by external sources, it will be more sensitive if the electrodes are widely separated and the applied voltage is high enough to produce many ionizing collisions as the initial carriers cross the gap. One of the most common applications of this charge multiplication, or avalanche effect, is the proportional counter, which produces a large current in response to relatively weak ionizing radiation. Although large, this current is proportional to the ionization rate G_0 so that the counter can discriminate between radiation of different ionizing strength, such as beta and alpha rays.

If the field is made too large, the avalanche will develop so much space charge that the applied field is distorted, reducing it to the point that the current saturates. If this happens, all pulses grow to about the same current level, and the counter, although extremely sensitive, can no longer discriminate among the various types of radiation. When operated in this mode the device is known as a Geiger counter.

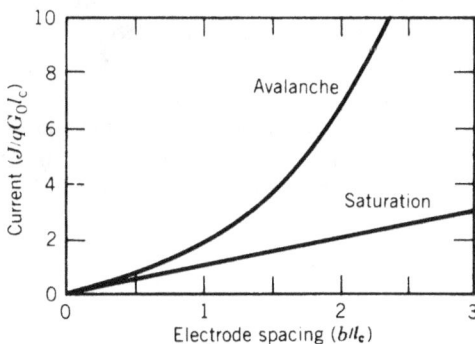

FIGURE 7.2.2. Current flow in the presence of avalanches.

7.3   CONDUCTION WITH MULTIPLE SPECIES
(Ion Bombardment Breakdown)

Summary

When more than one type of charge carrier is important, conservation equations for each type must be solved simultaneously to determine the net current flow, which is the superposition of each species' contributions. If many types of species are present, this technique goes over into a statistical theory of distributions. An example in which two species interact is the electrical breakdown caused by positive ion impact ionization at the electrodes.

Theory

The structure of the model for current flow when multiple species interact is a straightforward extension of the single-species model. If there are significant differences in charge, velocity, reactivity, or other characteristics of the various species, then similar ones are grouped together, and conservation equations are set up for each species in the form

$$\frac{\partial n_i}{\partial t} + \nabla \cdot \Gamma_i = G_i - R_i \tag{7.3.1}$$

Species with different charges must always be separated, because the current is given by an expression of the form

$$J = \sum_i q_i \Gamma_i \tag{7.3.2}$$

which requires individual determination of particle flux for each charge. In addition, the other terms in the conservation equation may call for additional groupings. For example, the negative charge may be carried by both negative ions and electrons, but the electrons will usually move much faster than the ions. The vast disparity in speed will usually require separate conservation equations for each of the two negative species. Finally, the generation and recombination terms may introduce new species, depending on the nature of these processes. In plasmas, for instance, ionization of neutral atoms is an important source of ions and electrons, and the density of these neutral atoms must also be included in the rate equations.

Often the number of separate species types becomes too large to handle. If so, the current summation is replaced by an integral over all possible values of some species parameter

$$J = \oint \frac{\partial}{\partial \xi} [q(\xi)\Gamma(\xi)] \, d\xi \tag{7.3.3}$$

In this equation ξ stands for the property of interest, which depends on the particular application. For electrons in a gas, ξ is the energy of the electron, while $\partial\Gamma(\xi)/\partial\xi$ might be related to the Maxwell–Boltzmann distribution. The actual distribution will depend on the solution of the rate equations under the particular conditions in the application. Distribution functions for electrons receive a great deal of attention in solid-state and plasma physics, but these techniques are just as valid for any group of charge species.

Example: Electrical Breakdown

The multiplication of the initial charge carrier density by the avalanche effect of collisional ionization causes a very large increase in the current flowing from the gap, but this does not mean that the gap has broken down. If the source of the ionization is removed, then there will no longer be the initial carriers to cause avalanches, and the current will drop to zero. In breakdown, however, the current will continue to flow even if the external source of ionization is removed.

For this to happen there must be some internal mechanism for producing the initial carriers. One of the common sources for such additional ionization is the bombardment of the electrodes by the heavy positive ions which result from the ionization in the insulating material. These ions move too slowly to ionize the tenuous medium through which they move, but when they strike the electrode they often have enough energy to release one or more electrons from the electrode surface. These new electrons are then accelerated away from the electrode and act just like the original charges which came from bulk ionization. The crucial prediction required of the model for such a process is the voltage at which breakdown will occur.

Application of the Theory

Since this is a surface effect, it changes only the boundary conditions. Assume that the electrodes are located at $x = -b$ and $x = 0$. The electron density between the electrodes still has the form

$$n_- = \frac{G_0}{\beta}(K_1 e^{-x/l_c} - 1) \tag{7.3.4}$$

given in Section 7.2. At $x = 0$, where the electrode is being bombarded, the electron flux is no longer zero. It is given in terms of the incoming positive ion flux as

$$|\Gamma_-(x = 0)| = |K_b \Gamma_+(x = 0)| \tag{7.3.5}$$

where K_b is the number of electrons emitted when an ion strikes the electrode. Since the total current is

$$J = J_- + J_+ \tag{7.3.6}$$

the positive and negative ion currents at $x = 0$ can be represented as

$$J_+(x = 0) = \frac{J}{1 + K_b} \tag{7.3.7}$$

$$J_-(x = 0) = \frac{K_b J}{1 + K_b} \tag{7.3.8}$$

This last equation also gives the electron density at the bombarded electrode as

$$n_-(x = 0) = \frac{J_-(x = 0)}{|\mu_-q_-|E} = \frac{K_bJ}{|\mu_-q_-|E(1 + K_b)} \qquad (7.3.9)$$

At this point the electron density is known at one electrode in terms of the current. This is sufficient to determine the constant K_1 in Eq. (7.3.5), giving the electron density expression as

$$n_-(x) = \frac{G_0}{\beta}(e^{-x/l_c} - 1) + \frac{K_bJ}{|\mu_-q_-|E(1 + K_b)}e^{-x/l_c} \qquad (7.3.10)$$

The total current can now be evaluated at any point, but the electrode at $x = -b$ is most convenient since the ion current vanishes there. The current is

$$J = J_-(x = -b) = n_-(x = -b)|\mu_-q_-|E$$

$$= \frac{|\mu_-q_-|EG_0}{\beta}(e^{b/l_c} - 1) + \frac{K_bJ}{1 + K_b}e^{b/l_c} \qquad (7.3.11)$$

Solving for J gives

$$J = \frac{|q_-|G_0l_c(e^{b/l_c} - 1)}{1 - [K_be^{b/l_c}/(1 + K_b)]} \qquad (7.3.12)$$

This current expression differs from the one obtained for the avalanche regime in one important respect. The denominator can be zero, indicating that the current flow becomes infinite along with the carrier density. This is the condition corresponding to breakdown. It indicates that enough electrons are generated by ion bombardment to enable the external ionization source to be turned off without cutting off the current flow. The condition for breakdown is given by

$$\frac{K_be^{b/l_c}}{1 + K_b} > 1 \qquad (7.3.13)$$

In addition to the electrode spacing, it contains the two parameters of the system l_c and K_b. This last parameter K_b is known as the second Townsend coefficient.

Discussion

The theory of this section was presented in general terms, but the application to the example was quite specific, requiring careful handling of the boundary conditions to get any solution. Unfortunately, this is often true for the complex modeling required when dealing with multiple species. In fact there are whole fields of specialization, such as plasma physics and solid-state physics, which are concerned almost entirely with solutions of the rate equations in specific media. Their methods become quite involved at times, but their basic goals are the same as those presented here, namely, to find the electric current in terms of species density and flux.

The method of solution used here is similar to that used earlier for space-charge flows, since it begins by assuming a uniform current density. Because

$$\nabla \cdot J = -\frac{\partial \rho}{\partial t}$$

this starting point is often helpful in the steady state. The desired result, of course, is usually an expression for current in terms of applied voltage, so we begin by assuming the answer and then working backward to find the voltage required to produce this current.

BIBLIOGRAPHY

Adamczewski, I., *Ionization, Conductivity and Breakdown in Dielectric Liquids*, Barnes and Noble, New York, 1969.

Atten, P., and J. C. Lacroix, Non-linear hydrodynamic stability of liquids subjected to unipolar injection, *J. Mec.*, **18**: 469–510 (1979).

Brown, S. C., *Introduction to Electric Discharges in Gases*, Wiley, New York, 1966.

Cobine, J. D., *Gaseous Conductors*, Dover, New York, 1958, Chap. 7.

Devins, J. C., The Physics of partial discharges in solid dielectrics, *IEEE Trans. Electr. Insul.*, **EI-19**: 475–496 (1984).

Felici, N. J., Electrostatics and hydrodynamics, *J. Electrostat.*, **4**: 119–129 (1977).

Felici, N. J., and J. C. Lacroix, Electroconvection in insulating liquids with special reference to uni- and bi-polar injection, *J. Electrostat.*, **5**: 135–144 (1978).

Gallo, C. F., Corona — A brief status report, *IEEE Trans. Ind. Appl.*, **IA-13**: 550–557 (1977).

Kok, J. A., *Electrical Breakdown of Insulating Liquids*, Interscience, New York, 1961.

Kuffel, E., and M. Abdullah, *High Voltage Engineering*, Pergamon, Oxford, 1970.

Lewis, T. J., The role of electrodes in conduction and breakdown phenomena in solid dielectrics, *IEEE Trans. Electr. Insul.*, **EI-19**: 210–216 (1984).

Loeh, L. B., *Basic Processes of Gaseous Electronics*, University of California Press, Berkeley, CA 1960.

Nassar, E., *Fundamentals of Gaseous Ionization and Plasma Electronics*, Wiley, New York, 1971, Chap. 8.

O'Dwyer, J. J., Breakdown in solid dielectrics, *IEEE Trans. Electr. Insul.*, **EI-17**: 484–492 (1982).

O'Dwyer, J. J., *The Theory of Electrical Conduction and Breakdown in Solid Dielectrics*, Clarendon, Oxford, 1973.

Schmidt, W. F., Electronic conduction processes in dielectric liquids, *IEEE Trans. Electr. Insul.*, **EI-19**: 389–418 (1984).

Sharbaugh, A. H., and P. K. Watson, Conduction and Breakdown in Liquid Dielectrics, *Prog. Dielectr.* **4**: 199–248 (1962).

Thomson, J. J., and G. P. Thomson, *Conduction of Electricity Through Gases*, Vol. II-2, Dover, New York, 1969.

Tobazeon, R., Electrohydronamic instabilities and electroconvection in the transient and A. C. regime of unipolar injection in insulating liquids: a review, *J. Electrostat.*, **15**: 359–384 (1984).

Yu, Tae-U, and G. M. Colver, Spark breakdown of particulate clouds: a new testing device, *IEEE-IAS Annu. Meet. Conf. Rec.*, *Chicago*, 1000–1009 (October 1984).

PROBLEMS

PROBLEM 1

A photocell is constructed from two electrodes of area A spaced a distance d apart. Light shines on the grounded electrode at $x = 0$, producing a photocurrent of electrons $J_-(x = 0) = J_0$. Electrons are also produced inside the cell by radiation and by collision, so that the total bulk generation rate is $G = G_0 + \beta n$. The electric field is strong enough so that diffusion, recombination, and space charge may be neglected.

a. Find the steady-state current if a positive voltage V is applied to the electrode at $x = d$.

b. Estimate this current, using the values

$$\mu = 10^{-4} \text{ m}^2/\text{V-s}, \qquad J_0 = 10^{-6} \text{ A/m}^2, \qquad A = 10^{-3} \text{ m}^2$$

$$G_0 = 10^8 \text{ m}^{-3} \text{ s}^{-1}, \qquad \beta = 100 \text{ s}^{-1}, \qquad V = 100 \text{ V}, \qquad d = 0.01 \text{ m}$$

PROBLEM 2

A smoke detector consists of two electrodes of area A spaced a distance d apart. An α-particle source maintains the positive ion density at n_0 near the grounded electrode at $x = 0$, but the bulk generation rate is $G = 0$. A large electric field pulls the ions across to the opposite electrode at $x = d$ so fast that diffusion, recombination, and space charge are negligible, and only the ion mobility μ_i is important.

a. What is the current I in terms of the voltage V on the electrode at $x = d$?

b. Smoke particles now enter the chamber and the ions attached to them, so that the ion density n_i is given by

$$\frac{\partial n_i}{\partial t} + \nabla \cdot \Gamma_i = -\beta n_i$$

and the charged smoke particle density by

$$\frac{\partial n_s}{\partial t} + \nabla \cdot \Gamma_i = +\beta n_i$$

Since the smoke particle mobility μ_s is usually less than the ion mobility, the current will initially be reduced and will then reach a steady value. What is this steady value?

c. Estimate the two currents, using the values

$$\mu_i = 10^{-4} \text{ m}^2/\text{V-s}, \qquad \mu_s = 10^{-5} \text{ m}^2/\text{V-s}, \qquad n_0 = 10^{14} \text{ m}^{-3}$$

$$V = 10 \text{ V}, \qquad d = 0.01 \text{ m}, \qquad A = 10^{-4} \text{ m}^2, \qquad \beta = 10 \text{ s}^{-1}$$

PROBLEM 3

In Section 7.3 breakdown was discussed with the assumption that ion bombardment of the cathode produces extra electrons. A faster process is the photoemis-
sion of electrons by light emitted from excited atoms left along the path of the
electron. Find the breakdown criterion for this process by following these steps:

a. Use conservation of electrons to find the general equation for the electron
    density between the electrodes. Assume that space charge is very small so
    that $E(x) = -V/d$. Recombination is very slow, and the generation is due
    to background radiation as well as collisions, so $G = G_0 + \beta n$.
b. Evaluate the integration constant in the general solution by assuming that
    the number of photoelectrons is proportional to the total number of elec-
    trons which exist between the plates, that is,

$$J_-(x = 0) = K_p \int_0^d n_-(x)\, dx$$

The constant K_p will depend on geometry, speed of electrons, and background gas density.

c. Combine these results to get the total current for the cell.
d. What is the breakdown condition?

PROBLEM 4

When the second ionization process in breakdown is caused by photoionization at
the cathode, the breakdown voltage can be raised if the light is absorbed before it
reaches the cathode. If the attenuation of the light is exponential, the boundary
condition for the electron current at the cathode becomes

$$J_-(x = 0) = K_p \int_0^l n(x)\, \exp(-K_a x)\, dx$$

Use the procedure outlined in Problem 3 to find the breakdown condition for this
case, assuming $G = G_0 + \beta n$ and $R = 0$.

CHAPTER

8

POLARIZATION

In conduction the charge carriers in the material are set into motion by an applied field, and this motion continues as long as the field is applied. In many materials, however, the charge carriers cannot continue to move indefinitely and may not be able to move at all in some circumstances. These restrictions on the charge motion give rise to the phenomena which are collectively referred to as polarization.

Polarization includes some of the oldest and some of the most useful aspects of electrostatics. The charge which can be stored on amber after rubbing it with fur furnished the name for electricity (from the Greek $\eta\lambda\epsilon\kappa\tau\rho\sigma\nu$, amber). Today, stored charges appear in numerous electrostatic devices and applications, all of which depend on the tendency of charges to resist continued motion in the same direction.

In this chapter the basic fields arising from charge separation are first derived and used to illustrate the construction of an artificial dielectric material. This example is linear, but the same fundamental approach is also valid for nonlinear dielectrics such as a ferroelectric. A related effect is the permanent polarization which results from immobile charges trapped in a material. In the last decade electret devices based on this effect have become quite prominent. One common application, the electret microphone, is discussed in more detail.

When the material holding the charges moves, the charges may move at different rates, and the polarization may change as a result. When this occurs the material is called piezoelectric. In some cases even thermal expansion can lead to motions which affect polarization, giving rise to pyroelectricity. This effect is used to provide thermal images for infrared television cameras.

Many of the common applications of polarization involve linear dielectrics in the sinusoidal steady state. This special case is illlustrated by the electronic resonance which forms the basis of microwave ovens.

8.1 POLARIZATION
(Artificial Linear Dielectrics)

Summary

Dielectrics are characterized by an internal charge arrangement which can be considered as a separation of equal numbers of positive and negative charges. Externally this rearrangement makes its presence known by an alteration in the charge on an electrode, compared to the value obtained in free space. The relation between the internal charge distribution and the apparent external behavior is exemplified by an artificial dielectric.

Theory

The basic measurable quantities in electrostatics are charge (or current) and electric field (or voltage). For example, if a known voltage v is applied to a parallel plate capacitor in a vacuum, a net charge q will flow to the upper plate, as shown in Figure 8.1.1, while the lower plate acquires the opposite charge. In the parallel plate geometry, the electric field between the plates is

$$E = \frac{v}{l} \tag{8.1.1}$$

From Gauss' law, the charge on the upper plate is

$$q = \iint D \cdot dA = \epsilon_0 EA \tag{8.1.2}$$

The relation between charge and voltage is

$$q = \frac{\epsilon_0 A}{l} v \equiv Cv \tag{8.1.3}$$

a linear relation which defines the capacitance between parallel electrodes in a vacuum. Alternatively, we could write the relation in terms of the field quanti-

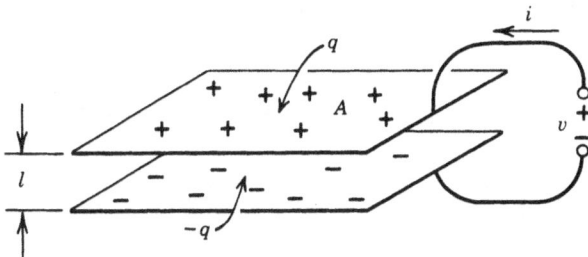

FIGURE 8.1.1. The relation between charge and voltage.

ties as

$$D\left(=\frac{q}{A}\right) = \epsilon_0 E \tag{8.1.4}$$

which is, again, a linear relation describing the permittivity of the vacuum.

These relations are strictly true only for a vacuum. If anything is placed between the electrodes, the fields will change, along with the charge on the electrodes. In practical application, a wide variety of materials may be placed between the electrodes, ranging from charged particles to crystalline solids. Each new material will affect the fields and terminal charge differently, but all of these new relations can be explained in terms of charges between the electrodes.

The electrical properties of most dielectric materials are determined by the separation of positive and negative charges within the dielectric materials. This separation can occur in many ways, but the easiest to understand is the separation into two thin layers of opposite charges as shown in Figure 8.1.2. These charges are the only material between the electrodes. Each layer has the same magnitude of charge, with opposite signs.

Since we know the magnitude and the location of all of the charges, we can solve for the fields in the regions above, below, and between the charge layers, using the techniques of Chapter 3. These fields are

$$E_a = E_b = \frac{v}{l} + \frac{\rho_s}{\epsilon_0}\frac{d}{l} \tag{8.1.5}$$

$$E_d = \frac{v}{l} - \frac{\rho_s}{\epsilon_0}\frac{l-d}{l} \tag{8.1.6}$$

The insertion of the charge layers *increases* the electric field adjacent to the upper electrode. From Gauss' law, this implies an increase in the charge on that

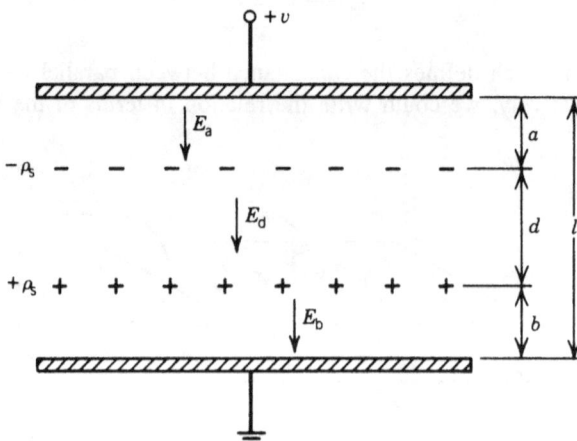

FIGURE 8.1.2. Charge layers between electrodes.

electrode, to

$$q = \epsilon_0 A E_a = \frac{\epsilon_0 A v}{l} + \frac{\rho_s \, dA}{l} \qquad (8.1.7)$$

The extra charge is proportional to the charge on each layer, and also to the separation between the layers. The product of charge and spacing is called the dipole moment.

If the charge magnitude and location were not specified, we could not calculate the extra charge, but we could still measure the relation between charge and voltage at the terminal; it would have the form

$$q = \frac{\epsilon_0 A v}{l} + q_p \qquad (8.1.8)$$

where the extra electrode charge q_p caused by the charge distribution between the electrodes is called the polarization charge.

The electric field between the electrodes is no longer uniform. It has been increased next to the electrodes but decreased between the charge layers. This internal nonuniformity is not apparent from terminal voltage measurements, however, since the voltage and spacing between electrodes is unchanged. The apparent or average electric field is still

$$E_{app} = \frac{v}{l} \qquad (8.1.9)$$

Thus, from the viewpoint of measurements made at the terminals, the introduction of the charges between the electrodes has not changed the apparent electric field that we are applying but it has changed the charge on the electrodes. The terminal relation can be rewritten as

$$q = \epsilon_0 A E_{app} + q_p \qquad (8.1.10)$$

or, using Gauss' law

$$D = \frac{q}{A} = \epsilon_0 E_{app} + \frac{q_p}{A} \qquad (8.1.11)$$

The last term occurs so often that it is defined separately as

$$P = \frac{q_p}{A} = \frac{\rho_s d}{l} \qquad (8.1.12)$$

where P is the polarization. The relation between D and E is thus

$$D = \epsilon_0 E_{app} + P \qquad (8.1.13)$$

This expression is valid for any number of charges, so long as the net charge is zero. With this constraint, the charge distribution will always be decomposed into pairs of opposite charges formed by separation of neutral entities. Materials in which this occurs are often called dielectrics; they can be solid, liquid, gaseous, or composed of individual particles.

It should be noted that the apparent E field is not the E field which actually exists next to the electrode. In fact, in our example, the electric field differs from the apparent field everywhere between the electrodes. Since the actual electric field inside a dielectric is rarely known precisely, only the apparent E field is normally used when discussing dielectrics, so that the influence of the material is expressed solely in terms of the polarization P. Thus, in writing the relation, the subscript is dropped with the understanding that the electric field is the average over the dielectric, or

$$D = \epsilon_0 E + P \qquad (8.1.14)$$

Example: Artificial Dielectrics

Most good insulators have relatively low dielectric constants. Often, however, there is a need for high capacitance in a small space, along with low resistance. This can be obtained by mixing the good insulator with conducting particles. A common formulation is metallic spheres dispersed throughout a potting compound such as epoxy, or a sandwich of insulating and conductive foils can be rolled up as in a lumped capacitor. In either case introduction of conducting elements between the electrodes increases the apparent capacitance (or permittivity) while using insulators with low dielectric constants. In designing such devices the apparent increase in dielectric constant can be estimated from the arrangement of the conductors.

Application of the Theory

Consider the effect of introducing a conducting foil into the space between two electrodes, as shown in Figure 8.1.3. Since the insert is conducting, the electric

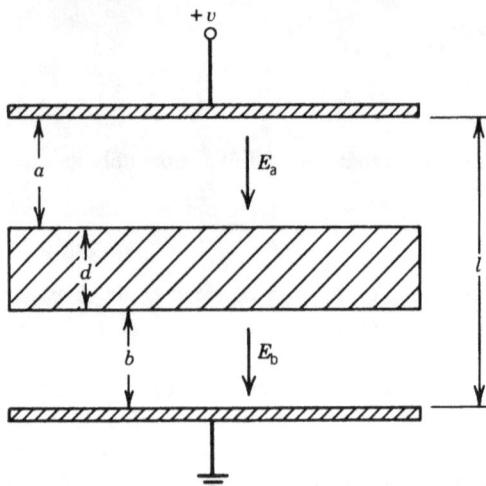

FIGURE 8.1.3. Polarization caused by a conducting foil.

field inside is $E_d = 0$, while outside the slab the fields are given by

$$E_a = E_b = \frac{v}{l - d} \tag{8.1.15}$$

The total charge on the upper electrode, by Gauss' law, is

$$q = \frac{\epsilon_0 A v}{l - d} \tag{8.1.16}$$

which can be rewritten in the polarization form

$$q = \frac{\epsilon_0 A v}{l} + \epsilon_0 A \frac{d}{l(l - d)} v \tag{8.1.17}$$

or, in terms of D and the apparent electric field E,

$$D = \frac{q}{A} = \epsilon_0 E + P \tag{8.1.18}$$

where the last term is the polarization given here by

$$P = \epsilon_0 \left(\frac{d}{l - d} \right) E \tag{8.1.19}$$

In this example the polarization is linearly proportional to the applied electric field, and the dielectric is termed *linear*. In a linear dielectric the polarization always has the form

$$P = \alpha E \tag{8.1.20}$$

The constant α is called the polarizability of the material. (A related quantity, atomic polarizability, is occasionally defined in terms of dipole moment as $p = \alpha E$.) This quantity can not be directly measured, but it is useful because most detailed theoretical models of dielectrics predict α. In the example under study here

$$\alpha = \frac{d\epsilon_0}{l - d} \tag{8.1.21}$$

Since both terms which contribute to D are proportional to E, they are often lumped together for convenience in the form

$$D = \epsilon_0 E + \alpha E \equiv \epsilon E \tag{8.1.22}$$

The material permittivity ϵ includes linear polarization. Most often, ϵ is expressed in terms of the permittivity of free space, ϵ_0, by the dielectric constant,

$$\kappa = \frac{\epsilon}{\epsilon_0} \tag{8.1.23}$$

For the example of an artificial dielectric, the apparent dielectric constant is

$$\kappa = \frac{l}{l - d} \tag{8.1.24}$$

Discussion

The model presented assumes that an infinite conducting sheet is interposed between the electrodes. The effective dielectric constant

$$\kappa = \frac{l}{l - d} = \frac{1}{1 - (d/l)} \tag{8.1.25}$$

does not depend on the position of the sheet but only on the fraction of the volume occupied by the sheet. The dielectric constant would be unchanged if the sheet were split into several smaller sheets. The apparent dielectric constant is plotted in Figure 8.1.4. For example, if 50% of the volume is filled with conducting inclusions, we expect the dielectric constant to double.

Discussion

Mixtures do not follow the volume fraction rule perfectly because the shapes of the conducting particles are usually not modeled well as flat sheets in alignment with the electrodes. The basic trend still holds true, however, with large increases in dielectric constant as the conductors fill the volume.

The opposite arrangement is also used when the low dielectric constant of a gas must be combined with the mechanical support of solids. In this case a molten insulating plastic is mixed with gas to form a foam. On cooling, the foam has sufficient strength to support the electrodes, but since the slightly conducting plastic occupies so little volume, the apparent dielectric constant is close to unity.

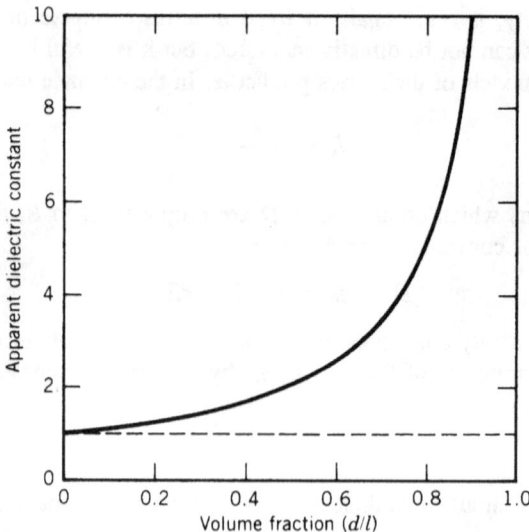

FIGURE 8.1.4. Apparent dielectric constant with conducting filler.

The polarization can be expressed in terms of individual dipoles as

$$P = \frac{\rho_s d}{l} = \frac{n_s}{l}(qd) \qquad (8.1.26)$$

where n_s is the number of dipoles per unit area. Each dipole has a moment $p = qd$ directed from minus to plus, so the polarization can also be written as a vector

$$\mathbf{P} = n_d \mathbf{p} \qquad (8.1.27)$$

where $n_d = n_s/l$ is the volume density of dipoles. This form is used later to calculate the dielectric properties of materials.

8.2 NONLINEAR POLARIZATION
(Dielectric Saturation)

Summary

In many materials polarization does not increase linearly with applied field, usually because the motion of the charges is subject to complicated forces or torques due to the surrounding medium. Nonlinear polarization can still be predicted in terms of basic models, as illustrated by the onset of polarization in a ferroelectric material.

Theory

Most nonlinear dielectrics contain more or less permanent dipoles which can be reoriented to align with an external field. This is strictly true with gases and is usually a good starting model for liquids and solids also. As the external field increases, more and more dipoles rotate, until finally all are aligned, giving the maximum polarization.

The alignment, or rotation, of each dipole is determined by the interplay between the torque exerted by the external field and the other torques which arise from interactions of the dipole with its medium. These interactions may be caused by thermal agitation, crystal structure, energy loss, or other mechanisms, and are often quite difficult to describe in fundamental terms. Their effect, however, is always the application of a torque to the dipole, so they are often treated in a phenomenonogical way by specifying a torque which will lead to the observed behavior.

The torque exerted by the external electric field on the dipole is usually related to the electrostatic forces on a model dipole consisting of two charges q_1 and q_2 separated by a distance d, as shown in Figure 8.2.1. For simplicity, the electric field is assumed to be in the x direction. The torque will occur regardless of any special relations among the charges and electric fields. Its value for the situation

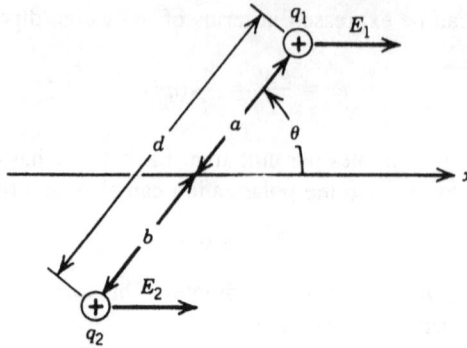

FIGURE 8.2.1. Model of an electric dipole.

shown in Figure 8.2.1 is

$$\tau = (-q_1 E_1 a + q_2 E_2 b) \sin \theta \qquad (8.2.1)$$

Obviously, the torque given by this expression depends on the distances a and b, and hence on the center chosen for the dipole. Mechanical systems are usually simplest to analyze if the center is chosen as the center of the applied force, which can be found by elementary mechanics procedures. In electrostatics problems this task is usually simplified by knowing that the charges are equal and opposite,

$$q_1 = -q_2 = q \qquad (8.2.2)$$

and by assuming that nonuniformities in the electric field have no effect, so that

$$E_1 = E_2 = E \qquad (8.2.3)$$

These assumptions give balanced forces with no net motion and a torque of

$$\tau = -qdE \sin \theta \qquad (8.2.4)$$

The dipole moment again appears ($p = qd$), allowing us to write

$$\tau = -pE \sin \theta \qquad (8.2.5)$$

Generalizing to three dimensions gives the torque on a dipole in a uniform field as

$$\tau = \mathbf{p} \times \mathbf{E} \qquad (8.2.6)$$

where the dipole moment vector is directed from negative to positive charge consistent with the definition of the polarization vector $\mathbf{P}$.

The relation between the external electric torque and the torques imposed by the medium determine the orientation of the individual dipoles. The material polarization $\mathbf{P}$ is the summation of all of these individual dipoles, so the final step involves an integration of the dipole moments over all possible orientations in the form

$$\mathbf{P} = \int d(n_d \mathbf{p}) \qquad (8.2.7)$$

Often the dipoles have a constant dipole moment and rotate in response to the field. The polarization in response to a z-directed electric field would be

$$P_z = \int |\mathbf{p}| \cos\theta \, \frac{dn_d}{d\Omega} \, d\Omega \qquad (8.2.8)$$

where θ is the angle with respect to the electric field and $dn_d/d\Omega$ is the number density of dipoles oriented within a given range of solid angle. For uniform orientation, the distribution is

$$\frac{dn_d}{d\Omega} = \frac{N_d}{4\pi}$$

so the polarization is

$$P_z = \int_{\psi=0}^{2\pi} \int_{\theta=0}^{\pi} \left(\frac{N_d |\mathbf{p}|}{4\pi} \cos\theta \right) \sin\theta \, d\theta \, d\psi = 0 \qquad (8.2.9)$$

A net polarization will arise only if the angular distribution of the dipoles is nonuniform.

Example: Ferroelectrics

One common type of nonlinear dielectric material shows little polarization until the applied field exceeds a threshold value, after which the polarization increases rapidly to a limiting value. Conceptually, this behavior might be expected if the dipoles were "stuck" in the surrounding media and could not be dislodged until the torque exceeded some minimum value τ_0. Once the critical torque is exceeded, however, the dipoles are free to turn, and they immediately line up with the field. This process involves energy wells and is quite similar to the promotion of electrons to the conduction band in semiconductors. We would like the find the dielectric response of such a material.

Application of the Theory

The first step in finding the polarization is the determination of the number of dipoles with a given orientation. This follows from the balance between the external electric torque and the torque supplied by the medium,

$$pE \sin\theta = \tau_m \qquad (8.2.10)$$

The material torque is not a simple function here, so it is best to show the solution of this equation graphically, as in Figure 8.2.2. In Figure 8.2.2a the electric torque is not large enough to overcome the well, even at 90° where it has its maximum. None of the dipoles will line up in this case. With a larger field, as in Figure 8.2.2b, the electric torque exceeds the material torque for dipoles which are oriented approximately perpendicular to the field. All of these dipoles will be reoriented to the field direction. The number reoriented will depend on whether

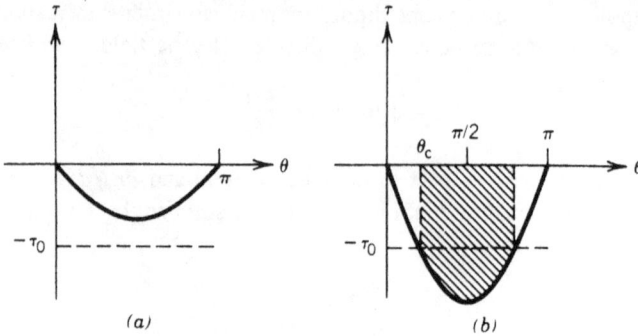

FIGURE 8.2.2. Dipole torque balance in a nonlinear dielectric.

their initial orientation was within the shaded portion of the figure, defined in terms of a critical angle given by

$$\theta_c = \arcsin \frac{\tau_0}{pE} \qquad (8.2.11)$$

The critical angle can exist only if the electric field is large enough to make the argument of the arcsine no greater than unity, so no realignment takes place at all below a critical field given by

$$E_c = \frac{\tau_0}{p} \qquad (8.2.12)$$

If the dipoles are initially distributed uniformly in all directions, their net dipole moment will vanish, so below the critical field the polarization is zero, as shown in Figure 8.2.3. When the critical field is exceeded, a fraction of the dipoles given by $\cos \theta_c$ will be aligned with the field, giving a polarization integral of

$$P = N_d p \sqrt{1 - \left(\frac{\tau_0}{pE}\right)^2} \qquad (8.2.13)$$

This response is also shown in Figure 8.2.3 for applied fields in excess of the critical value. As expected, the polarization shows a rapid increase above E_c, which levels off when most of the dipoles have broken loose from their wells.

Discussion

The curve presented here describes the polarization as the electric field increases from zero. In many devices of this type a similar process occurs as the electric field decreases, leading to a different nonlinear curve. This depends, of course, on the nature of the material torques in the new orientation aligned with the field. Because a great many torque relations are possible in a solid medium, a great variety of polarization relations are found in practice.

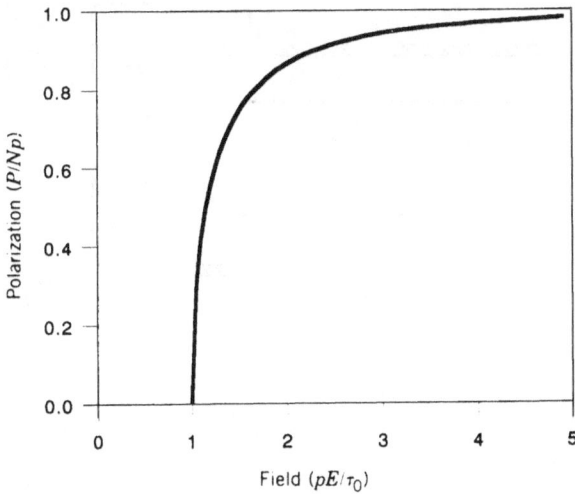

FIGURE 8.2.3. Polarization response of a ferroelectric material.

In the example a very simple model was found to be adequate to explain the shape of a typical ferroelectric polarization curve. This is not a proof that the model actually describes the molecular processes, of course, but it may represent a good starting point. Usually such simple models are developed further to explain other effects, such as temperature dependence, by adding elements of statistical mechanics, quantum mechanics, and crystal structure. These refinements are beyond the scope of this book, and for many materials they are still in the process of development.

8.3 PERMANENT POLARIZATION
(Electret Microphones)

Summary

In some dielectrics charges can be positioned and then locked in place, giving a permanent dipole moment (or charge separation) on a macroscopic scale. These materials, such as electrets, offer a portable source of dipole charge and can be used in ways similar to permanent magnets. One common example is the electret microphone, whose output is calculated here.

Theory

An electret is a dielectric material with a permanent dipole moment or charge separation. There are a number of ways of producing such a charge separation. One of the earliest, sketched in Figure 8.3.1, uses an electric field to pull charges through

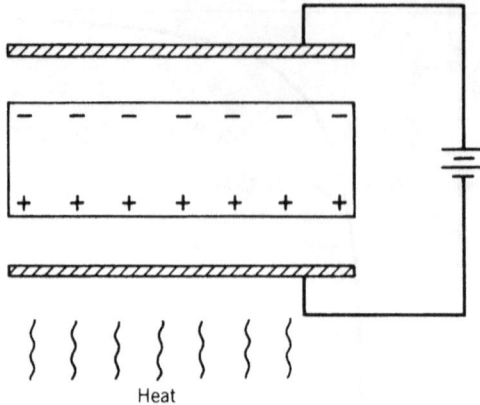

FIGURE 8.3.1.   Thermal electret formation.

a wax which has been softened by heating. After the desired charge separation
occurs, the wax is cooled and the charges are "frozen" into their new positions.
Turning off the external field causes no further change since the conductivity of
the cool wax is very small. Other materials, such as Mylar (PET), can form electrets by this process. Thermal formation can also involve the reorientation of
dipoles, which are frozen into place in much the same manner as the mobile
charges were. In all of these thermal charging methods, the charge in the electret
is the opposite of that on the nearby electrode. These are called heterocharge
electrets.

   A second type of electret (homocharge) is formed when like charges are
deposited from the electrode, as in the corona-charged electret shown in Figure 8.3.2. With this method the charge is directly deposited from the electrode.
No heating is required since no charge motion in the dielectric is required. Similar
methods employing other charge sources, such as sparks or electron beams, are
also widely used. The electron beam method is somewhat more flexible than the
others since the position of the charges is controlled by the energy of the electrons. For typical materials a 10-keV electron will stop within 1 μm of the surface, whereas a 30-keV electron might penetrate 10 μm. Since devices usually

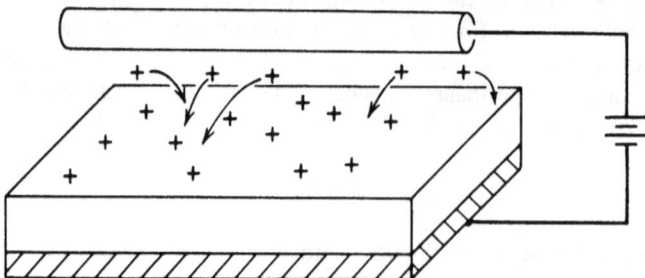

FIGURE 8.3.2.   Formation of an electret by corona.

work better if the charge is close to the surface, low energy electrons are generally employed.

With modern insulating materials any of these methods are capable of forming an electret strong enough to break down electrically, so the application of electrets is limited more by the electric field distribution than by the method of manufacture. The fields depend on the charge position, of course, and can usually be found by the methods covered earlier.

Example: Electret Microphone

Most high fidelity microphones now incorporate an electret transducer element in an arrangement similar to that shown in Figure 8.3.3. Sound waves strike the electret membrane, moving it with respect to the metal plate. This motion varies both the charge and voltage on the plate, either of which can be used as an output signal.

Application of the Theory

The fields in the regions above and below the charge layer have constant values E_a and E_b and satisfy the boundary conditions

$$E_a a + E_b x = v \tag{8.3.1}$$

$$\epsilon E_a - \epsilon_0 E_b = \rho_s \tag{8.3.2}$$

The solution for the fields is

$$E_a = \frac{\epsilon_0 V + x\rho_s}{a\epsilon_0 + x\epsilon} \tag{8.3.3}$$

$$E_b = \frac{\epsilon V - a\rho_s}{a\epsilon_0 + x\epsilon} \tag{8.3.4}$$

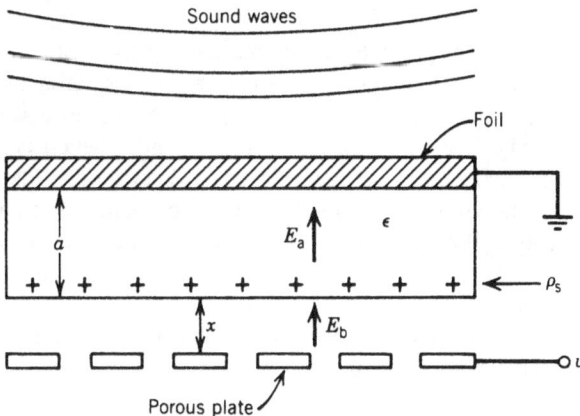

FIGURE 8.3.3. An electret microphone.

When the electret is first formed, the lower electrode is usually far away, whereas the upper electrode is usually permanently attached. In this situation the distance x is much greater than the electret thickness a, so that the field exists only inside the electret. This field is

$$E_a = \frac{\rho_s}{\epsilon} \qquad (8.3.5)$$

The dielectric constant of electrets is greater than that of air, so the electric field is weaker than it would be in air. In addition, the breakdown field is usually larger in dielectrics, so a greater surface charge can be applied to a foil-backed electrode than could be sustained in air without breakdown.

If the microphone electrodes are held at the same potential $v = 0$, the charge induced on the bottom electrode is given by

$$\frac{q}{A} = D = \epsilon_0 E_b = -\left(\kappa + \frac{x}{a}\right)^{-1} \rho_s \qquad (8.3.6)$$

As the position x changes in response to acoustic waves striking the membrane, the charge changes, leading to a current flow.

For a typical 1 mil (25 μm) thick Teflon ($\kappa = 2.1$) electret microphone the surface charge might be $\rho_s \approx 50 \ \mu C/m^2$, and the air gap might be 20 μm at rest. The electric field at the output electrode is then $E \approx 2.1$ MV/m, close to air breakdown. The output depends on the amplitude of the diaphragm displacement, which depends on the loudness of the sound and on the stiffness of the mounting. This dependence is nonlinear, so the motion should be kept very small if distortion is to be avoided.

8.4 PIEZO- AND PYROELECTRICITY
(Infrared Television)

Theory

Most textbook treatments of piezoelectricity and the closely related pyroelectricity are couched in terms of tensors and crystal classes, which tends to obscure the basic simplicity of the effect. In addition, this approach implies that piezoelectric effects can occur only in well-defined crystalline solids, which is not true. To present a simpler introduction to the piezoelectric effect, we consider the electrostatics of the model shown in Figure 8.4.1. The two capacitor plates are shown as short circuited, so that all of the effects are due to the charge layer, located at x. This assumption can be relaxed if desired, but neglecting the external voltage allows the piezoelectric effect to be exhibited more easily.

The surface charge ρ_{se} induced on the upper electrode is obtained by the solution of the electrostatic field equations, as

$$\rho_{se} = \frac{x}{d} \rho_{sx} \qquad (8.4.1)$$

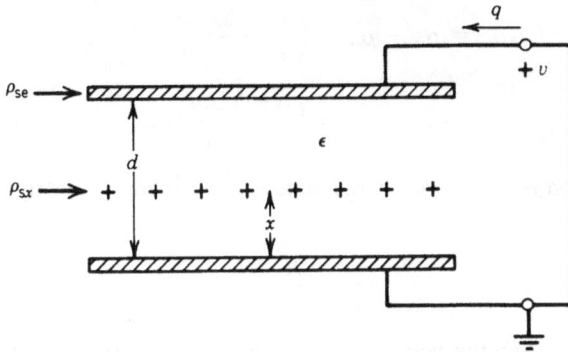

FIGURE 8.4.1. Basic model of piezoelectricity.

assuming a uniform permittivity. This basic arrangement has been studied before, but the positions of the charge layer and electrodes have always been held constant. In the basic piezoelectric effect, however, the electrodes move relative to each other, which may induce a change in the charge on both electrodes. At the same time, the internal charge layer may also be disturbed by the motion, since the charges are mechanically connected through the surrounding medium to the electrodes. After the electrodes have been moved to $d + \Delta d$, the charge layer will have moved to $x + \Delta x$, leaving the new charge ρ'_{se} on the upper electrode as

$$\rho'_{se} = \rho_{sx}\left(\frac{x + \Delta x}{d + \Delta d}\right) = \rho_{se}\frac{(1 + \Delta x/x)}{(1 + \Delta d/d)} \tag{8.4.2}$$

There is no a priori relation between Δd and Δx; it must be determined by the mechanical constraints. The simplest assumption is that of uniform expansion in which

$$\frac{\Delta x}{x} = \frac{\Delta d}{d} \tag{8.4.3}$$

In this case,

$$\rho'_{se} = \rho_{se} \tag{8.4.4}$$

and there is no variation in electrode charge caused by the motion. This situation arises if the charges are "tied" to the surrounding medium so that they move together.

In many situations, particularly with solids, the charges are not free to move with the background medium. This can occur in crystalline solids if the surrounding crystal structure causes the charges to move sideways under extension, or when molecules have been aligned by mechanical stresses. In this case,

$$\frac{\Delta x}{x} \neq \frac{\Delta d}{d} \tag{8.4.5}$$

and the charge on the electrode changes by

$$\Delta\rho_{se} = \rho'_{se} - \rho_{se} = \left(\frac{1 + \Delta x/x}{1 + \Delta d/d} - 1\right)\rho_{se} \qquad (8.4.6)$$

$$\approx \left(\frac{\Delta x}{x} - \frac{\Delta d}{d}\right)\rho_{se}$$

If the compression is small, the charge motion is likely to be linearly related to the electrode motion

$$\frac{\Delta x}{x} = \frac{K'\Delta d}{d} \qquad (8.4.7)$$

so the relation between the polarization and the displacement has the form

$$\Delta P = \Delta\rho_{se} \approx (K' - 1)\frac{\Delta d}{d} = K_{pz}\frac{\Delta d}{d} \qquad (8.4.8)$$

Since the electrodes are short circuited, the electrode charge is due solely to polarization P. The constant K_{pz} is a piezoelectric coefficient.

Most piezoelectric materials have more than one type of charge inside. Often there are dipoles associated with the molecular structure, and there may be additional layers of trapped charges. Since the electrostatic fields are linear, each of these additional charges contributes an induced charge with a similar relation to compression. Summing them yields a net result of the same form as Eq. (8.4.8). These additional charges do not change the form of the piezoelectric equation, but of course the value of the coefficient will depend on the details of charge magnitude and position and on the relation of charge motion to the motion of the medium.

Heating a material can also lead to relative charge motion and polarization. The mechanism of the effect is closely related to piezoelectricity, with thermal expansion replacing external forces in distorting the material. If the material has the appropriate anisotropy, it does not matter what causes the distortion, since the induced polarization is defined only in terms of the resulting charge movements. Heating of the material usually leads to expansion, which can then cause polarization by means of piezoelectric effects. This combination of thermal expansion and piezoelectricity is called pyroelectricity.

Conceptually this chain of events is represented by the piezoelectric coefficient and thermal expansion coefficient as

$$\Delta P = K_{pz}\left(\frac{\Delta d}{d}\right) = K_{pz}K_{th}\Delta T \equiv K_{py}\Delta T \qquad (8.4.9)$$

The factor K_{py} is called a pyroelectric coefficient. A typical value for a widely used pyroelectric material (triglycine sulfate or TGS) is $K_{py} = 200 \ \mu C/m^2 K$, but values can range from 0.001 to 17,000 $\mu C/m^2 K$.

The foregoing discussion is somewhat superficial since the stresses in these anisotropic materials are tensor quantities whose interrelations are much more complicated than Eq. (8.4.9). Unfortunately, a careful treatment on the introductory level is more likely to confuse than instruct and is not attempted here. This simple phenomenological approach is still useful in practice, since most pyroelectric materials are characterized by a single pyroelectric coefficient relating charge and temperature.

Example: Infrared Television

Most optical sensors rely on the generation of charge carriers by incoming radiation. The photogenerated carriers can then generate a current directly, as in photodiodes, or indirectly, as in photoconductors. This type of sensor works well at short wavelengths because the high energy photons can easily produce ionization. Infrared light, however, has relatively low energy and is rarely capable of causing ionization. Some other means of detection must be used.

Pyroelectricity is well suited to detection of infrared radiation since this wavelength is very effective in heating materials. It has been used in a number of devices, including the infrared television cameras which can see in the dark and are widely used to detect energy losses from buildings.

Application of the Theory

The single-parameter model gives a polarization charge which is proportional to the temperature of the material. The electronic instrumentation used to process this signal is responsive to current, so that the output is a function of the polarization current

$$J_p = \frac{dP}{dt} = K_{py}\frac{dT}{dt} \tag{8.4.10}$$

or

$$i = K_{py}A\frac{dT}{dt} \tag{8.4.11}$$

for the output from a cell with area A. The cell gives an output only when the temperature is changing, which is not what is desired in an infrared camera. A steady image in a device like this will display a uniform blank picture. Moving the camera will give a recognizable image, but a better approach is to chop the incoming light with a shutter, so that individual cells experience a constantly varying temperature.

The pyroelectric material has a leakage resistivity and permittivity in addition to the pyroelectric interaction, so the pyroelectric current is injected into the equivalent RC circuit shown in Figure 8.4.2. The output voltage is given by solution of the equation

$$C\frac{dv}{dt} + \frac{v}{R} = i_{py} = K_{py}A\frac{dT}{dt} \tag{8.4.12}$$

Clearly, the output voltage will vanish unless the temperature is changing. If the

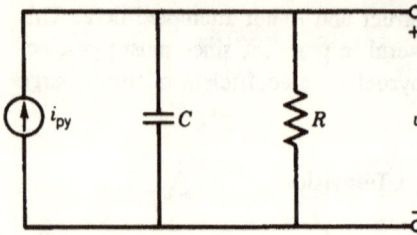

FIGURE 8.4.2. Equivalent circuit of pyroelectric detector.

temperature varies sinusoidally,

$$T = \text{Re}[\Delta T e^{j\omega t}] \qquad (8.4.13)$$

The magnitude of the voltage response is

$$|\Delta v| = K_{py} A R\, \Delta T \frac{\omega}{\sqrt{1 + (\omega RC)^2}} \qquad (8.4.14)$$

The frequency response is shown in Figure 8.4.3. The light should be chopped at a frequency greater than

$$f_c \simeq \frac{1}{2\pi RC} = \frac{\sigma}{2\pi\epsilon} \qquad (8.4.15)$$

to ensure high output. Typical values for triglycine sulfate (TGS) are

$$\sigma = \frac{1}{1.7} \times 10^{-10}\ \Omega\text{-m} \qquad (8.4.16)$$

$$\epsilon = 35\epsilon_0 \qquad (8.4.17)$$

so the thermal image must be chopped much faster than approximately 0.03 Hz to image correctly.

The output voltage under these conditions is

$$\Delta v \simeq K_{py} A R\, \Delta T \approx \frac{K_{py} d\, \Delta T}{\sigma} \qquad (8.4.18)$$

for a layer of material of thickness d. For TGS with a typical thickness of $d =$

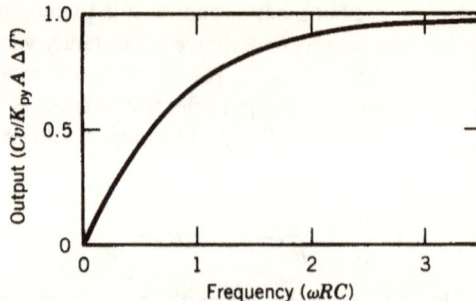

FIGURE 8.4.3. Frequency response of a pyroelectric detector.

20 μm, a 1-K temperature variation gives an output voltage of

$$|\Delta v| = \frac{K_{py}d}{\sigma} \Delta T \simeq 68 \text{ V} \qquad (8.4.19)$$

8.5   DIELECTRICS IN TIME-VARYING FIELDS
(Microwave Ovens)

Introduction

So far, the current flow and the polarization have resulted from essentially static fields. In many applications of dielectrics the applied fields and resulting currents are changing in time. When this occurs it is sometimes difficult to separate conduction processes from polarization effects. In the sinusoidal steady state, for instance, either permittivity or conductivity can represent charge motion with equal accuracy, but either one may have a complex frequency dependence. One example is the resonance in the conductivity of water which makes the microwave oven possible.

Theory

Everything in this chapter is based on the application of charge conservation to the device under study. Typically, we are interested in finding the current flowing to an electrode along with a change in the net charge on the electrode, as illustrated in Figure 8.5.1. Conservation of charge for the volume enclosing the electrode gives

$$i = \iint \mathbf{J} \cdot d\mathbf{A} + \frac{\partial q}{\partial t} \qquad (8.5.1)$$

or

$$i = \iint \left(\mathbf{J} + \frac{\partial \mathbf{D}}{\partial t} \right) \cdot d\mathbf{A} \qquad (8.5.2)$$

using Gauss' law, and assuming that the enclosing surface is fixed in space.

From the external circuit, where i is measured, it is impossible to separate the conduction current J from the displacement current $\partial D/\partial t$ without some additional information about motion of charges in the vicinity of the electrode. For example, a constant current in the electrode lead could result from free charge motion at constant velocity toward the electrode, or from a constantly increasing E field in the presence of fixed charges. Since detailed information of this sort is often lacking in practice, the two contributions are often lumped together to give a total current,

$$J_T = J + \frac{\partial D}{\partial t} \qquad (8.5.3)$$

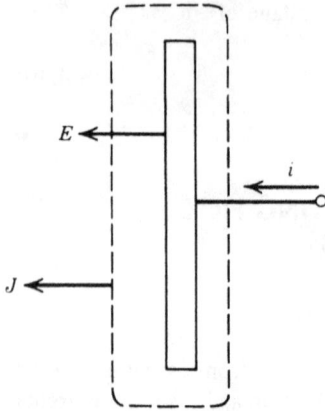

FIGURE 8.5.1. Charge flow at an electrode.

The last few sections have discussed only some of the relations which charac-terize J and $\partial D/\partial t$ in practical devices, but even this limited selection is too broad to draw any general conclusions about the total current which will flow. Instead, we discuss an important special case based on the sinusoidal steady state, in which an electric field of the form

$$E = \text{Re}[\hat{E}e^{j\omega t}] \tag{8.5.4}$$

is applied to the dielectric. We assume that both the current and the polarization which result from this electric field are linear functions of the field so that

$$J = \text{Re}[\hat{J}e^{j\omega t}] \tag{8.5.5}$$

and

$$D = \text{Re}[\hat{D}e^{j\omega t}] \tag{8.5.6}$$

are both sinusoids at the same frequency as the applied field. For example, in the simple case of an ohmic conductor with linear polarization,

$$\hat{J} = \sigma\hat{E} \tag{8.5.7}$$

$$\hat{D} = \epsilon\hat{E} \tag{8.5.8}$$

the total current is given by

$$\hat{J}_{\text{T}} = (\sigma + j\omega\epsilon)\hat{E} \tag{8.5.9}$$

Note that this relation has the same form as Ohm's law. In fact, it is often called the complex Ohm's law and defines a complex conductivity by

$$\hat{J} = \sigma^*\hat{E} \tag{8.5.10}$$

where

$$\sigma^* = \sigma + j\omega\epsilon \tag{8.5.11}$$

It is not a very general relation, since it is only valid in the sinusoidal steady state, but this occurs so often in engineering work that it deserves special study.

Since the equation is complex, there is no a priori reason for selecting the conductivity as the prime factor. We could just as easily define a complex permittivity by

$$\hat{D} = j\omega\epsilon^*\hat{E} \tag{8.5.12}$$

where

$$\epsilon^* = \epsilon - \frac{j\sigma}{\omega} \tag{8.5.13}$$

In fact, this expression is often preferable in dealing with dielectrics since the conduction current is usually much smaller than the displacement current.

In both of these forms it has become customary to speak of the real and imaginary parts of the complex conductivity, which are defined as

$$\sigma^* = \sigma_r + j\sigma_i \tag{8.5.14}$$

where $\sigma_r = \sigma$ and $\sigma_i = \omega\epsilon$ in our example. Similarly, the complex permittivity is written as

$$\epsilon^* = \epsilon_r + j\epsilon_i \tag{8.5.15}$$

where $\epsilon_r = \epsilon$ and $\epsilon_i = \sigma/\omega$ in the ohmic example. This is merely a matter of redefinition and would not be worth mentioning except that the real and imaginary parts of both conductivity and permittivity are related to each other since they describe the same phenomenon.

Example: Microwave Ovens

Microwave ovens are possible because the motion of electrons in water does not follow the simple mobility law which is often assumed in conductors. Instead, the electrons are subject simultaneously to restoring forces and inertial effects, while the drag imparted by the medium is relatively light. The combination of mass and restoring force leads to natural harmonic oscillations, which can reach large amplitudes at the proper electric field frequency. These large motions, together with the drag forces, heat the surrounding medium so that energy is transformed from the electric field to electron motion and thence into heat.

Application of Theory

The total current is still given by

$$J_T = nqu + \frac{\partial D}{\partial t} \tag{8.5.16}$$

but the velocity is now determined by

$$m\frac{d^2x}{dt^2} = -Kx + qE \tag{8.5.17}$$

where Kx is the restoring force on the electron. With sinusoidal excitation,

$$\hat{x} = \frac{q}{K - m\omega^2}\hat{E}$$ (8.5.18)

and

$$\hat{u} = j\omega\hat{x} = \frac{j\omega q}{K - m\omega^2}\hat{E}$$ (8.5.19)

which gives the total current as

$$\hat{J}_T = \left(\frac{j\omega nq^2}{K - m\omega^2} + j\omega\epsilon\right)\hat{E}$$ (8.5.20)

The complex current has only an imaginary part so that the situation is best expressed in terms of the complex permittivity, which has only a real part

$$\epsilon' = \epsilon + \frac{nq^2/m}{\omega_0^2 - \omega^2}$$ (8.5.21)

where the resonant frequency is

$$\omega_0 = \sqrt{\frac{K}{m}}$$ (8.5.22)

This permittivity is plotted in Figure 8.5.2.

In the absence of damping, the permittivity reaches extremely large values in the vicinity of the resonant frequency. This resonance is associated with large displacements and high velocities of the electrons. If drag is present, these high velocities lead to heating of the material at that frequency, which is the basis of the microwave oven.

Discussion

By completely neglecting the drag forces which give rise to mobility, we developed a model which predicts a very large displacement current (or permittivity) in

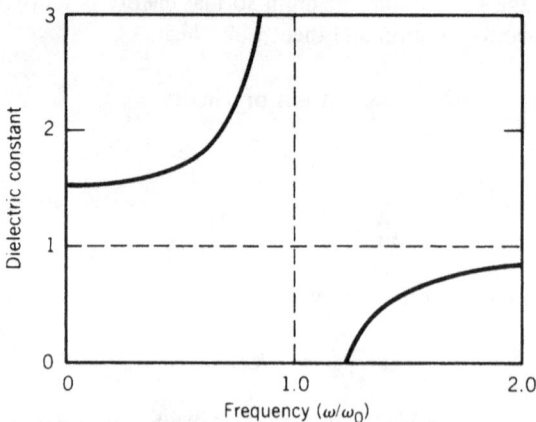

FIGURE 8.5.2. Frequency response for microwave resonance.

the vicinity of the resonant frequency. These resonances are quite common in materials and often form the basis for identifying unknown samples, since the resonant frequency depends on the effective mass and restoring force, which vary from one material to another.

When drag forces are included in the model, the electron motion is still large at the resonant frequency, but there is also a resistive component, which dissipates the energy coming from the electric field by means of electron motion. This dissipation of electric energy by water also plays an important role in radar, where rain can absorb the waves, thus shielding whatever lies beyond.

BIBLIOGRAPHY

Böttcher, C. J., *Theory of Electric Polarization*, Elsevier, Amsterdam, 1952.

Brown, J., Artificial dielectrics, *Prog. Dielectr.*, **2:** 193–225 (1960).

Burfoot, J. C., and G. W. Taylor, *Polar Dielectrics and Their Applications*, University of California Press, Berkeley, 1978.

Cady, W. G., *Piezoelectricity*, Vols. 1 and 2, Dover, New York, 1964.

Daniel, V. V., *Dielectric Relaxation*, Academic, New York, 1967.

Grindlay, J., *An Introduction to the Phenomenological Theory of Ferroelectricity*, Pergamon, Oxford, 1970.

Gross, B., *Charge Storage in Solid Dielectrics*, Elsevier, Amsterdam, 1964.

Jefeminko, O. D., Long-lasting electrization and electrets, in *Electrostatics and Its Applications*, A. D. Moore, Ed., Wiley, New York, 1973, Chap. 6.

Ketani, M. A., Piezoelectric power generation, in *Direct Energy Conversion*, Addison-Wesley, Reading, MA, 1970, Chap. 10.

Lang, S. B., *Sourcebook of Pyroelectricity*, Gordon and Breach, New York, 1974.

Liebert, L., *Liquid Crystals*, Academic, New York, 1978.

Lines, M. E., and A. M. Glass, *Principles and Applications of Ferroelectrics and Related Materials*, Clarendon, Oxford, 1977.

Nelson, D. F., *Electric, Optic, and Acoustic Interactions in Dielectrics*, Wiley, New York, 1979.

Nussbaum, A., *Electromagnetic and Quantum Properties of Matter*, Prentice-Hall, Englewood Cliffs, NJ, 1966, Chap. 5.

O'Konski, C. T., *Molecular Electro-optics*, Vol. 1, Marcel Dekker, New York, 1976.

Pohl, H. A., Non-uniform field effects: dielectrophoresis, in *Electrostatics and its Applications*, A. D. Moore, Ed., Wiley, New York, 1973, Chap. 4.

Sessler, G. M., Ed., *Electrets*, vol. 1, Laplacian Press, Morgan Hill, CA, 1998.

Von Hippel, A. R., *Dielectrics and Waves*, Wiley, New York, 1954.

Von Hippel, A. R., *Dielectric Materials and Applications*, Wiley, New York, 1954.

Woodson, H. H., and J. R. Melcher, *Electromechanical Dynamics*, Vol. 3, Wiley, 1968, Chap. 11.

PROBLEMS

PROBLEM 1   (NATURAL SCIENCE)

All materials exhibit electronic polarization in which the nucleus and the surrounding electron cloud move relative to each other under the influence of an external field. Assume the electronic charge $-q$ is uniformly distributed over a sphere of radius d and the nucleus is displaced a distance x from the center.

a. Find the displacement where the electrostatic force from the external field E_0 balances the electron field.

b. What is the polarizability of this molecule?

PROBLEM 2   (COMPUTER PERIPHERALS)

A material (like a liquid crystal) is made of dipoles with dipole moment p which can rotate against a restoring torque $K\theta$. Normally the dipole is horizontal ($\theta = 0$), but when an electric field $\mathbf{E} = E_0\mathbf{i}_y$ is applied, it turns the dipole to a new angle θ.

a. Find the induced dipole moment in the y direction in terms of the applied electric field.

b. If there are N of these dipoles per unit volume, what is the polarization for this material?

PROBLEM 3   (ELECTRONICS)

There are N electrons per unit volume in a solid. A typical electron is held near $x = 0$ by a springlike restoring force $F = -kx$.

a. Find the dielectric constant for this solid.

b. For large deflections of the electron, the force is nonlinear, with the variation shown in Figure P.8.3. If the electric field E_0 increases from zero to $2F_0/q$, find $P(E)$ and plot it versus E_0.

c. If the electric field now decreases from $2F_0/q$ to zero, find and plot $P(E)$.

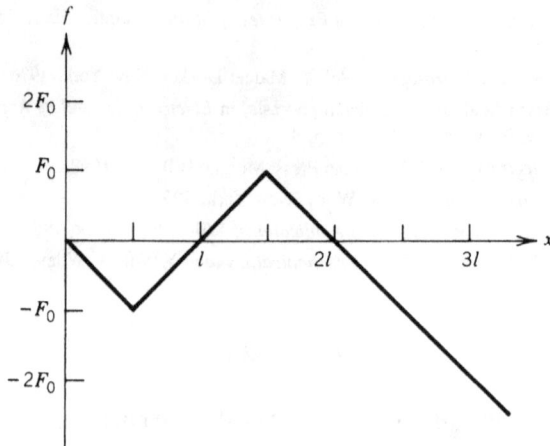

FIGURE 8.P.3.   Restoring force in a nonlinear dielectric.

PROBLEM 4 (INSTRUMENTATION)

An electrophorus is formed from a sheet of Teflon of thickness d (= 100 μm) and dielectric constant κ (= 3). One side is attached to a grounded metallic foil, while the other side is free. A layer of charge ρ_s (= 10 μC/m^2) is injected into the polymer at a distance a (= 40 μm) from the foil.

a. Assuming that any other electrodes are far away, find the electric field everywhere.

b. In using the electrophorus, a second grounded electrode is placed on the free side. Find the charge induced on this electrode.

c. This electrode is now disconnected from ground and removed from the electrophorus. How much charge does it have? How much is left inside the electrophorus?

PROBLEM 5 (INSTRUMENTATION)

Piezoelectric transducers usually drive a load, and must supply a voltage. Consider the device of Figure 8.4.1 with the shorting connection removed.

a. Find the terminal relation (q versus v) for this device by superposing the charges induced by the internal charge layer and the external voltage.

b. If the transducer is attached to a resistor R in series with a bias voltage v_b, find the equation for the output voltage in terms of d and x.

PROBLEM 6 (INSTRUMENTATION)

An elastic material between electrodes contains a layer of dipoles with charge ρ_s and separation $d < 1/2$. The dipoles are rigid and cannot change length or charge. The material thickness is l ($\gg d$), and its area is A.

a. Find the terminal relation $q = q(v, \rho_s)$ for this device.

b. Initially the terminals are short circuited. The device is then compressed to a total thickness of $l/2$ with the terminals open circuited. Find the change in voltage at the terminals.

PROBLEM 7

Find the complex permittivity for the microwave heating example of Section 8.5 if there is an additional force $-B(dx/dt)$ due to drag on the electron by the surrounding medium.

CHAPTER

9

CONTINUUM FORCE
DENSITIES

The motions of individual charged particles can be very well described by Coulomb's force, but the net force of an electric field on a material containing a large
number of charges is not so easily determined. Part of the problem arises from the
bookkeeping needed to keep track of all of the particles and the forces acting on
them if the particles are very numerous. The best approach involves an averaging
over a small volume to give the force density, which depends on net charge density, not on individual charges. Force densities can be derived for Coulomb's
forces, as well as polarization forces, and the densities can be transformed into
tensor forms which allow further simplifications in the arithmetic.

There is a further problem connected with collections of charges which is often
overlooked. The methods to be described give the net force on all of the charges
in the volume, but this number may be meaningless if the charges have no collective effect. Usually we tacitly assume that the electric force on each charge is immediately transmitted to the medium by means of drag or restoring forces. If so,
the net force is effectively exerted on the medium. If the charges can slip through
the medium unhindered, however, the force just goes into acceleration of the
charges. This ambiguity must be resolved before the results of force calculations
can be applied.

The simplest force density involves the Coulomb force acting on a collection of
charges, which gives rise to the basic electrostatic force density. This force, when
coupled by ion drag to a liquid, drives a fluid pump which is used to cool transformers. The more complicated polarization force, which acts on dipoles in a
nonuniform field, has been suggested for fuel orientation inside spacecraft in a
weightless environment.

Both of these force densities can be converted to an alternate form involving tensors, which allows the net force to be calculated in terms of the electric field at the boundaries of the region. This approach is often simpler when the internal field distribution is complex. It is illustrated for an electric field meter used in an underwater power cable.

9.1 FORCE DENSITIES
(Ion Drag Pumps)

Summary

Electrostatic forces on materials, rather than on isolated charges, depend on coupling between the charge and the material. Once this coupling has been achieved, the force acting on a small volume of the material can be expressed in terms of a modified Coulomb force. This expression is used to estimate the pressure head generated by an ion drag pump recently developed to cool high voltage transformers.

Theory

The basic electrostatic force is that on a very small charged particle q given by

$$f = qE \qquad (9.1.1)$$

This form has been useful in following the motions of individual charges, but it can not be used when there are large number of charges acting collectively, owing to the difficulty of keeping track of the individual charges. An alternate formulation involving force density is usually a better choice in this situation.

A force density is the net force acting on a volume containing the charges. Consider a small region of space containing several individual charge carriers, as shown in Figure 9.1.1. Each individual charge experiences a force given by Eq. (9.1.1). The electric field E consists of contributions from the externally applied field E_0 as well as contributions due to the electrostatic fields between pairs of particles within the volume. The net force acting on all of the charges within the volume is obtained by summing all of the forces acting on each charge within

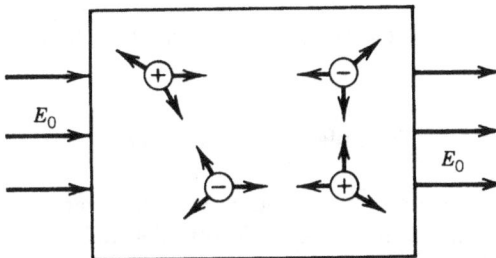

FIGURE 9.1.1. A small volume containing charges.

the volume. Since the interaction forces between charge pairs are all equal and opposite, they cancel in the summation, and the net force is given by

$$f_{net} = \sum_i (n_i d\mathcal{V}) q_i E_0$$

where E_0 is the external (or average) electric field over the volume and n is the number of charges per unit volume. Thus the force on a unit volume of charges is given by

$$\mathbf{F} = \lim_{d\mathcal{V} \to 0} \left(\frac{\mathbf{f}}{d\mathcal{V}} \right) = nq\,\mathbf{E}_0 = \rho\mathbf{E}_0 \tag{9.1.3}$$

This equation is a straightforward extension of the electrostatic force equation, but there is an additional assumption which is often implicit in its application. Strictly speaking, this is the net force on a collection of charges, but the equation is often used as if it were the net force on the material containing the charges. For example, if the charges are in air, the force density is assumed to act on the air, as well as on the charges. Occasionally this assumption is false, so it is important to understand when it can be made.

When a force is applied to a charged particle, its first effect is to accelerate the particle, and all of the force is directed into this acceleration. As the particle begins to move, it will interact with the surrounding molecules, transferring some of its momentum. At this time part of the force (on the average) is accelerating the particle, and part is acting on the surrounding material. Eventually, the charges reach an equilibrium where the electrostatic force just balances the force imposed by the medium. At this point the electrostatic force, in effect, is acting directly on the medium, and the force on the charge collection can be considered as the force on the medium. This interaction of drag and momentum has been discussed in Chapter 4, where the controlling parameter was found to be the momentum decay time, given by t_m. If this time is short compared to the mechanical times in the system, the force density can be considered to act directly on the medium.

If the total force, rather than its density, is required, the force density can be integrated in the usual manner over the volume of interest to give

$$f = \iiint \rho(\mathbf{r})E(\mathbf{r})\,d\mathcal{V} \tag{9.1.4}$$

Of course, the variation of the charge density and the electric field must first be determined.

Example: Ion Drag Pump

In high voltage power equipment, the current-carrying components must be cooled to carry away the heat caused by ohmic losses. Generally, fluid cooling is more effective than solid conduction and is used whenever high powers are encountered. The fluids are most effective when they make direct contact with the high voltage conductors, but this requires that they be electrical insulators such as oils or gases. The coolant must also circulate to carry away the heat, so some means of pumping by thermal convection or mechanical pumps is generally included.

An alternate method, which is more effective than thermal convection and simpler than mechanical pumping, is based on the force transmitted to the cooling fluid by charges in an electric field. This electrohydrodynamic (EHD) pumping has been known for some time, but practical implementation in a transformer was reported only recently (Sharbaugh, 1983).

The most important quantity for pumping is the net force applied to the fluid over the pump. This can be calculated by using the methods of this section, once the charge density and electric field inside the pump are known.

Application of the Theory

A model of the pump is shown in Figure 9.1.2. The region between the electrodes contains a uniform charge density ρ_0 which is generated at the left electrode and collected at the right electrode. The potential difference between the electrodes is given by $v = V_0$. In practice, the space charge is usually nonuniform, and the electric field is distorted correspondingly. These distortions can be handled by considering space-charge limitations to the current flow, as in Chapter 5, but their inclusion at this point would just add to the complexity of the force density without giving any additional insight into the method.

The electric field inside the pump due to the electrodes and the internal space charge is

$$E = \frac{V_0}{d} + \frac{\rho_0}{\epsilon}\left(x - \frac{d}{2}\right) \qquad (9.1.5)$$

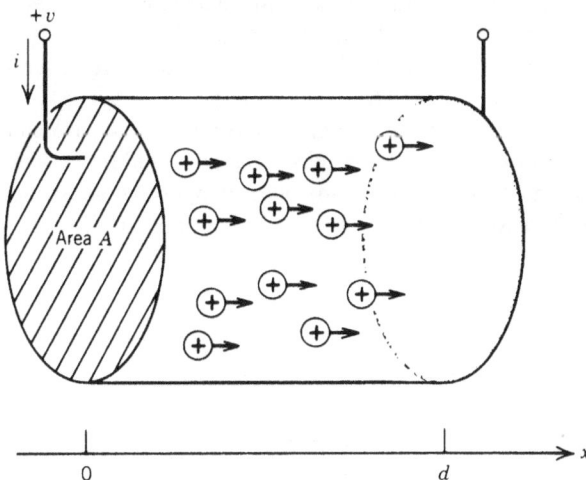

FIGURE 9.1.2. An EHD ion drag pump.

and the force density is

$$F = \rho E = \frac{\rho_0 V_0}{d} + \frac{\rho_0^2}{\epsilon}\left(x - \frac{d}{2}\right) \qquad (9.1.6)$$

Although the force density varies throughout the pump, the net force, which is of primary concern, is easier to interpret. It is given by

$$f = \int F \, d\mathcal{V} = \rho_0 A V_0 \qquad (9.1.7)$$

In effect, it is just the net charge in the pump multiplied by the electric field that would be applied if space-charge effects were negligible. In practice, the variation in force throughout the pump has an important bearing on the pump's output because the nonuniform pressure leads to internal stirring of the fluid, called electro-convection, which reduces the output. For the ion pump developed at General Electric (Sharbaugh, 1983), the applied voltage was on the order of $V_0 = 20$ kV. The maximum charge will be less than that needed to cancel the applied field, or

$$\rho_0 \leq \frac{2\epsilon V_0}{d^2} \qquad (9.1.8)$$

from Eq. (9.1.5). With $\epsilon = 2.2\epsilon_0$ and a spacing of $d = 5$ mm, the charge density $\rho_0 \leq 31$ mC/m^3. Thus the pressure f/A for the pump should be on the order of 623 N/m^2 under optimum conditions. This is sufficient to pump the oil at a few centimeters per second, which significantly increases the heat dissipation in the transformer. (Measured values were about one-half of this estimate).

Discussion

The force density is a simple generalization of the Coulomb force and finds wide application in continuum systems such as gases and liquids. It should be kept in mind that it tacitly assumes that the motion of the charges, averaged over some appropriate time, corresponds to equilibrium between the drag of the medium and the electrostatic force on the particle.

Its use is often part of an analysis that includes simultaneous determination of electric field distribution, charge distribution, and also fluid flows which depend in part on the force density. This can become quite complicated in practice, and much design work is done assuming relatively simple solutions for the flow, charge, and electric fields.

9.2   POLARIZATION FORCES
(Spacecraft Fuel Management)

Summary

Until now we have been finding the force on a particle which had a single charge, which could be taken as a point charge. In many applications the charge on a particle frequently has a more complicated distribution, which also affects the motion. The next step in complexity is the dipole charge, which influences the motion of many dielectrics, such as the liquid fuels used in spacecraft.

Theory

Just as a point charge is the simplest representation of a unipolar distribution, the discrete electric dipole is the simplest form which shows the essential features of dipole forces. In this model two point charges of different value q_1, q_2 are separated by a constant distance d, as shown in Figure 9.2.1. An electric field which varies in space exerts forces on the two charges.

If the charges were independent, each would move in response to the local electric field. Crucial to the dipole model, however, is the assumption that the two charges are somehow attached to each other, so that they must both move together. For example, the dipole may represent a molecule, such as water vapor in air, in which the positive and negative charges are held apart by the molecular structure but neither can leave the molecule because the ionization potential is too great. In another example, the dipole could exist on a drop of liquid in which the charges are induced by an external electric field, a situation which often occurs in raindrops.

Since the two charges are tied together, the net force on the combination is given by

$$f = q_1 E(x_1) + q_2 E(x_2) \tag{9.2.1}$$

This expression is generally correct, but it contains two charges and two values of electric field, which makes it somewhat difficult to draw any general conclusions. A common approach to a problem of double variables is to recast the variables in terms of average and difference values, so that

$$q_1 = q_{av} - \delta q \tag{9.2.2}$$

$$q_2 = q_{av} + \delta q \tag{9.2.3}$$

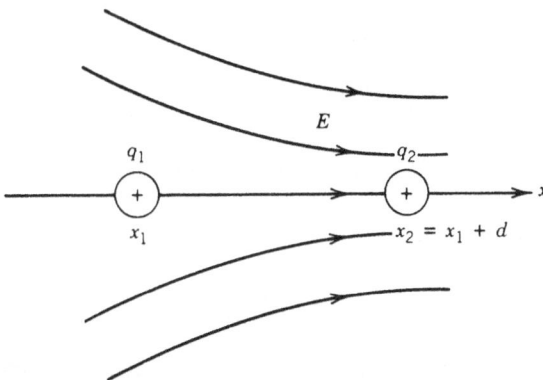

FIGURE 9.2.1. Dipole in an electric field.

and

$$E(x_1) = E_{av} - \delta E \qquad (9.2.4)$$

$$E(x_2) = E_{av} + \delta E \qquad (9.2.5)$$

Substituting these expressions into the net force equation yields

$$f = 2q_{av}E_{av} + 2(\delta q)(\delta E) \qquad (9.2.6)$$

This simplifies the net force to two terms which depend on average values and on difference values, respectively.

The average term

$$2q_{av}E_{av} = (q_1 + q_2)E_{av} \qquad (9.2.7)$$

gives the net force on the charge pair in terms of the total (or net) charge, as we would expect. Usually, when discussing dipole forces, the net charge is assumed to vanish,

$$q_2 = -q_1 = q \qquad (9.2.8)$$

and this force term is neglected. In practice, especially with macroscopic dipoles, the net charge remains and must be included in the force equations. The net charge force, when it occurs, is usually the dominant force, so its neglect should always be carefully justified.

The other term depends on charge difference and field difference. If either is lacking, this force will vanish. With the charges given by Eq. (9.2.8), the charge difference

$$\delta q = \frac{q_2 - q_1}{2} = q \qquad (9.2.9)$$

is always present, so the electric fields at the two charge positions must differ for the force to exist. If $E(x_2) > E(x_1)$, for example, the positive charge experiences a greater force, dragging the negative charge along, in spite of the opposing force. Often the electric fields do not differ much, and an approximate expression based on the Taylor series approximation gives an accurate value of δE. In this case

$$E(x_2) \simeq E(x_{av}) + \left.\frac{\partial E}{\partial x}\right|_{x_{av}}\left(\frac{d}{2}\right) + \cdots \qquad (9.2.10)$$

and

$$\delta E = E_+ - E_- \simeq d\left.\frac{\partial E}{\partial x}\right|_{x_{av}} \qquad (9.2.11)$$

neglecting all higher-order terms. This gives the dipole force term as

$$f = [qd]\frac{\partial E}{\partial x} \qquad (9.2.12)$$

The derivative can be evaluated anywhere in the vicinity of the dipole if the linear approximation is valid, so the subscript has been dropped. The term in the brackets is the dipole moment,

$$p = qd \qquad (9.2.13)$$

In the model of Figure 9.2.1, the distance d represents the horizontal component of the charge separation, which is parallel to the electric field. In terms of vector components, this can be written

$$f_x = p_x \frac{\partial E_x}{\partial x} \qquad (9.2.14)$$

where $p_x = qd_x$. Generalizing to field variations in other directions gives the x component of force as

$$f_x = p_x \frac{\partial E_x}{\partial x} + p_y \frac{\partial E_x}{\partial y} + p_z \frac{\partial E_x}{\partial z} = (\mathbf{p} \cdot \nabla)E_x \qquad (9.2.15)$$

where $\mathbf{p}$ is a dipole moment vector which is directed from the negative to the positive charge of the dipole. The total vector force on the dipole is

$$\mathbf{f} = (\mathbf{p} \cdot \nabla)\mathbf{E} \qquad (9.2.16)$$

Just as with the charged particles of Section 9.1, the force on a collection of electric dipoles can be obtained by adding the contributions of each individual dipole. This process leads to a force density

$$\mathbf{F} = (\mathbf{P} \cdot \nabla)\mathbf{E} \qquad (9.2.17)$$

where the polarization vector is just the total of all the individual dipoles per unit volume,

$$\mathbf{P} = n\mathbf{p} \qquad (9.2.18)$$

With the same assumptions about equilibrium between material interactions and applied force, the dipole force density can also be considered as acting on the volume of material containing the dipoles.

In many materials the polarization is a linear function of the applied electric field of the form

$$\mathbf{P} = \epsilon_0(\kappa - 1)\mathbf{E} \qquad (9.2.19)$$

so for linear isotropic dielectrics the polarization force density has the form

$$\mathbf{F}_p = \epsilon_0(\kappa - 1)(\mathbf{E} \cdot \nabla)\mathbf{E} = \tfrac{1}{2}\epsilon_0(\kappa - 1)\nabla(E^2) \qquad (9.2.20)$$

Example: Fuel Management in Spacecraft

In space there is no gravity to pull liquid fuel to the bottom of a tank. Instead, the fuel may slosh around, upsetting the balance of the spacecraft and cutting off the fuel supply if it floats away from the fuel line inlet. Several methods for controlling the fuel have been suggested, including the use of polarization forces. The

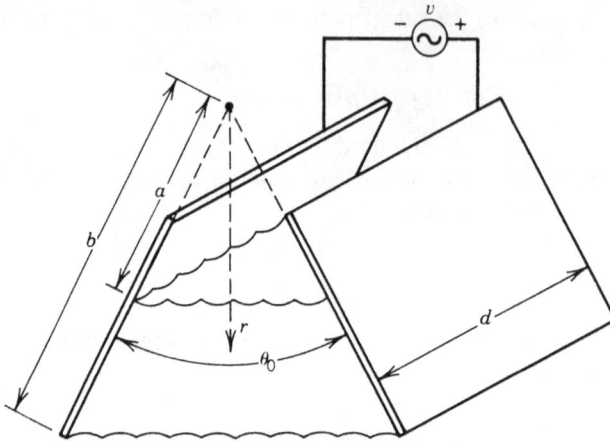

FIGURE 9.2.2. Polarization fuel management.

basic configuration for polarization control is shown in Figure 9.2.2. The fuel is contained between two electrodes which are arranged radially, with an angle θ_0 between them. They extend from $r = a$ to $r = b$ with a potential difference v. In practice, the applied voltage must be alternating, since the basic force on charges in fluids is usually much stronger than the dipole force, and dc components will mask the polarization effect which is desired here. The period of the applied ac voltage should be much shorter than the charge relaxation time of the fuel.

The success of this method depends on the magnitude of polarization force which can be generated. This force is calculated in the following section.

Application of the Theory

For this example, we assume

$$v = V_0 \cos \omega t \tag{9.2.21}$$

With cylindrical geometry, the electric field between the electrodes is purely azimuthal with a magnitude

$$E_\theta = \frac{V_0 \cos \omega t}{\theta_0 r} \tag{9.2.22}$$

The liquid fuel is a linear dielectric, so Eq. (9.2.20) can be used to find the force density as

$$\mathbf{F} = \frac{1}{2} \epsilon_0(\kappa - 1)\nabla(E^2) = -\frac{\epsilon_0(\kappa - 1)V_0^2}{\theta_0^2 r^3} \cos^2 \omega t \, \mathbf{i}_r \tag{9.2.23}$$

The total force on the volume of fuel between the electrodes is given by integration as

$$\mathbf{f} = \int_{z=0}^{d} \int_{\theta=0}^{\theta_0} \int_{r=a}^{b} \mathbf{F} r \, dr \, d\theta \, dz \; = \; \frac{-\epsilon_0 d(\kappa - 1)V_0^2 \cos^2 \omega t}{\theta_0} \left[\frac{1}{a} - \frac{1}{b} \right] \mathbf{i}_r,$$

$$(9.2.24)$$

For a typical fuel tank

$$\kappa = 2.5, \qquad d = 1 \text{ m}, \qquad V_0 = 50 \text{ kV}$$

$$a = 1 \text{ cm}, \qquad b = 1 \text{ m}, \qquad \theta_0 = 1 \text{ rad}$$

the peak force on the fuel is given by $F = 3.3$ N. This is far less than its weight on the earth (approximately 4000 N or 408 kg), but in space it is more than sufficient to overcome the other, weaker forces that remain when gravity is gone.

Discussion

The calculation presented here represents only a first try at designing a fuel management system based on polarization forces since it completely neglects the fluid nature of the fuel. In practice there are a great many additional motions which may occur in a fluid, including separation into individual blobs, which may render the containment system unreliable. All of these must be considered in the design of a practical device.

Although polarization forces are often considered weak, there are a great many contexts where they are far more effective than Coulomb forces since polarization forces can act directly on an uncharged mass. This area, called dielectrophoresis, has found applications in biology, sorting, and many other areas.

The form of the polarization force used here is only one of many ways in which this force can be expressed. Some of these forms are superficially dissimilar and have caused some concern among workers in the field. For most practical problems all the different formulations give the same results if they are properly used (Melcher, 1981, Chap. 3).

9.3   THE MAXWELL STRESS TENSOR
(Electric Field Meter)

Summary

When dealing with complex charge distributions, it is often difficult to find the forces which affect the motion of macroscopic objects by directly evaluating the Coulomb force density expression. An alternate method of calculating forces, called the Maxwell stress tensor, can often be used to find the forces in terms of the electric fields at the surface of a volume enclosing the system. This is illustrated by an electric field meter for use inside a power cable.

Theory

The net electrostatic force can always be obtained in principle by adding or integrating the force density, but it is often simpler in practice to use a different tool for force calculations, called the Maxwell stress tensor. This approach is similar in many ways to the use of electric potential rather than electric fields, since it makes use of a single entity, the stress tensor, which can be differentiated to give the individual components of the electric force on the region of interest.

The method is based on the force density obtained in Section 9.1,

$$\mathbf{F} = \rho\mathbf{E} \tag{9.3.1}$$

To use this expression to find the force acting on a particular collection of charges, we must know both the charge density and the electric field throughout the volume. If we make use of Gauss' law, the equation can be rewritten solely in terms of electric fields as

$$\mathbf{F} = (\nabla \cdot \mathbf{D})\mathbf{E} \tag{9.3.2}$$

The equations which follow can become complicated if all of the terms are displayed at each step, so we deal with only one component of the force density until the nature of the method is clear. Later, we set up simpler ways of writing our results. If we are looking for the x component of the force density, it may be written from Eq. (9.3.2) as

$$F_x = E_x\frac{\partial D_x}{\partial x} + E_x\frac{\partial D_y}{\partial y} + E_x\frac{\partial D_z}{\partial z} \tag{9.3.3}$$

where all of the terms are displayed explicitly. If we assume that the permittivity ϵ is independent of position, the first term in this expression can be rewritten as

$$E_x\frac{\partial(\epsilon E_x)}{\partial x} = \frac{\partial}{\partial x}\left(\frac{1}{2}\epsilon E_x^2\right) \tag{9.3.4}$$

The second term can be rewritten using the chain rule as

$$E_x\frac{\partial}{\partial y}(\epsilon E_y) = \frac{\partial}{\partial y}(\epsilon E_x E_y) - \epsilon E_y\frac{\partial E_x}{\partial y} \tag{9.3.5}$$

The first term on the right in this equation is the derivative of a scalar quantity similar to Eq. (9.3.4). The second term is not a perfect derivative, but if we use the electrostatic equation,

$$\nabla \times \mathbf{E} = 0 \tag{9.3.6}$$

we can make the substitution

$$\frac{\partial E_x}{\partial y} = \frac{\partial E_y}{\partial x} \tag{9.3.7}$$

so that the second term in Eq. (9.3.5) becomes

$$-\epsilon E_y \frac{\partial E_x}{\partial y} = -\epsilon E_y \frac{\partial E_y}{\partial x} = \frac{\partial}{\partial x}\left(-\frac{1}{2}\epsilon E_y^2\right) \tag{9.3.8}$$

which is also a perfect derivative.

The same process can be carried out on the last term in the force expression of Eq. (9.3.3) so that the force density in the x direction can be written in terms of derivatives as

$$F_x = \frac{\partial}{\partial x}\left[\frac{\epsilon}{2}(E_x^2 - E_y^2 - E_z^2)\right] + \frac{\partial}{\partial y}(\epsilon E_x E_y) + \frac{\partial}{\partial z}(\epsilon E_x E_z) \tag{9.3.9}$$

or as

$$F_x = \frac{\partial T_{xx}}{\partial x} + \frac{\partial T_{xy}}{\partial y} + \frac{\partial T_{xz}}{\partial z} \tag{9.3.10}$$

where the terms T_{ij} are defined by

$$T_{xx} = \frac{\epsilon}{2}(E_x^2 - E_y^2 - E_z^2) \tag{9.3.11}$$

$$T_{xy} = \epsilon E_x E_y \tag{9.3.12}$$

$$T_{xz} = \epsilon E_x E_z \tag{9.3.13}$$

The first subscript indicates the component of the force which we are seeking, whereas the second subscript indicates the derivative which must be performed on that term. From the form of the force expression, Eq. (9.3.10), it is clear that the force density is equivalent to the divergence of a vector given by

$$\mathbf{T}_x = (T_{xx})\mathbf{i}_x + (T_{xy})\mathbf{i}_y + (T_{xz})\mathbf{i}_z \tag{9.3.14}$$

or

$$F_x = \nabla \cdot \mathbf{T}_x$$

At this point, the force density in the x direction can be calculated by evaluating the derivatives in Eq. (9.3.10), but this does not really represent an improvement over the situation we started with in Eq. (9.3.2). In both of these expressions the same types of terms must be evaluated over the volume of interest. The advantage of the new formulation is that, because it gives the force in terms of a divergence of a vector, we can make use of the divergence theorem to calculate the net force over a large volume.

$$f_x = \iiint F_x d = \iiint (\nabla \cdot \mathbf{T}_x)\, d\mathcal{V} = \oiint \mathbf{T}_x \cdot \mathbf{n}\, dA \tag{9.3.15}$$

Since the force is given in terms of the electric fields at the surface, we do not need to know the electric field over the entire volume, which means that in many cases detailed solution of the field equations can be avoided. This is especially likely to be true for the complicated distributions of charges which result when conducting surfaces are connected to voltage sources.

So far, we have considered only the force in the x direction. Similar transformations can be used for the force in the other directions, so that for any direction x_i the force can be expressed as

$$F_i = \sum_{j=1}^{3} \frac{\partial T_{ij}}{\partial x_j} \tag{9.3.16}$$

In three-dimensional space, the indices i and j can take on the values $1, 2, 3$ corresponding to the three principal directions. The entire set of values for T_{ij} is

$$T_{ij} = \begin{bmatrix} \frac{\epsilon}{2}(E_x^2 - E_y^2 - E_z^2) & \epsilon E_x E_y & \epsilon E_x E_z \\ \epsilon E_y E_x & \frac{\epsilon}{2}(E_y^2 - E_x^2 - E_z^2) & \epsilon E_y E_z \\ \epsilon E_z E_x & \epsilon E_z E_y & \frac{\epsilon}{2}(E_z^2 - E_x^2 - E_y^2) \end{bmatrix} \tag{9.3.17}$$

This array is known as the Maxwell stress tensor; each of its rows can be considered as a vector whose divergence gives the force density due to electrostatic forces.

In working with tensors there are several notations which reduce the amount of writing needed. One is the summation convention, which is useful in expressions like

$$F_i = \sum_{j=1}^{3} \frac{\partial T_{ij}}{\partial x_j} \tag{9.3.18}$$

Instead of writing the summation sign explicitly, we assume that the expression should be summed from 1 to 3 whenever a subscript is repeated. Using the summation convention, the expression of Eq. (9.3.18) could be rewritten

$$F_i = \frac{\partial T_{ij}}{\partial x_j} \tag{9.3.19}$$

A second notational convention, the Kronecker delta, is often used when writing the components of the stress tensor. This function is defined as

$$\delta_{ij} = \begin{cases} 1 & \text{if } i = j \\ 0 & \text{if } i \neq j \end{cases} \tag{9.3.20}$$

By using the Kronecker delta function, all of the stress components of Eq. (9.3.17) may be expressed as

$$T_{ij} = \epsilon E_i E_j - \frac{\epsilon}{2}(E_k E_k)\delta_{ij} \tag{9.3.21}$$

Note that the repeated index k implies that

$$E_k E_k = E_1^2 + E_2^2 + E_3^2 = E_x^2 + E_y^2 + E_z^2 \tag{9.3.22}$$

by the summation convention.

Example: E Field Measurement

As an example of the use of the Maxwell stress tensor, let us study a method for measuring the electric field strength inside an electric power cable. This method was developed recently in connection with the Skagerrak cable which connects the power systems of Norway and Denmark (Nyberg, 1979). The Skagerrak cable is operated with dc voltages, and, since the apparent conductivity of the insulation varies with temperature and time in normal operation, the electric field strength inside the cable will tend to vary more than it would be with an ac cable.

To measure the electric field strength under various operating conditions, a portion of the outer conductor of the cable was replaced by a thin metallic membrane. A rectangular model of this device is shown in Figure 9.3.1. When the conductor is energized, charges appear on the two electrodes and attract each other. If the charges stay on the electrodes rather than entering the insulator, the force is transmitted to the membrane, causing it to deflect, which gives an indication of the field strength.

Application of Theory

The magnitude of the electric force acting on the membrane can be determined by using the Maxwell stress tensor. Consider the electrode with the embedded membrane as shown in Figure 9.3.2. If we want to find the force acting on any area of the membrane, we can integrate the Maxwell stress tensor over a surface which encloses the membrane, as shown in Figure 9.3.2. Because the field is uniform, it has only a single component in the insulator,

$$E_x = E_0, \qquad E_y = E_z = 0 \tag{9.3.23}$$

and the stress tensor at all points on the inside surface reduces to

$$T = \begin{bmatrix} \frac{1}{2}\epsilon E_0^2 & 0 & 0 \\ 0 & -\frac{1}{2}\epsilon E_0^2 & 0 \\ 0 & 0 & -\frac{1}{2}\epsilon E_0^2 \end{bmatrix} \tag{9.3.24}$$

Outside the cable, of course, the electric field, and hence the stress tensor, vanish.

The main concern in this device is the electric force acting normal to the membrane, or f_x. It is given by the area integral as

$$f_x = F_1 = \oiint T_{1j} n_j \, dA \tag{9.3.25}$$

There are four sides to the area shown in Figure 9.3.2; all of them must be included in performing the integral. The first side gives no contribution since the stress tensor vanishes outside the cable. The second side is defined by its normal vectors as

$$n_x = n_1 = 0, \qquad n_y = n_2 = 1 \tag{9.3.26}$$

Since only one of the components of the normal vector is nonzero, the area inte-

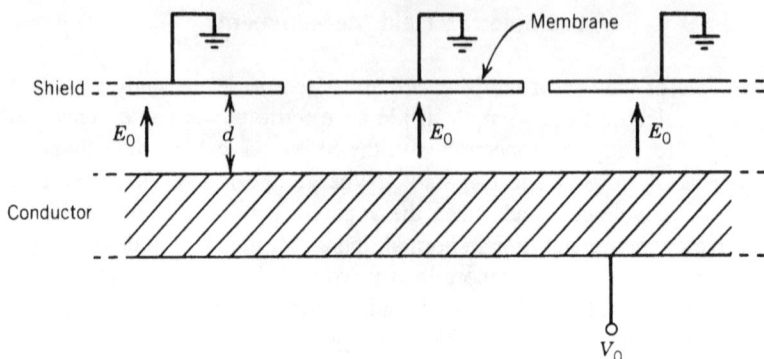

FIGURE 9.3.1. Flexible membrane in an electrode.

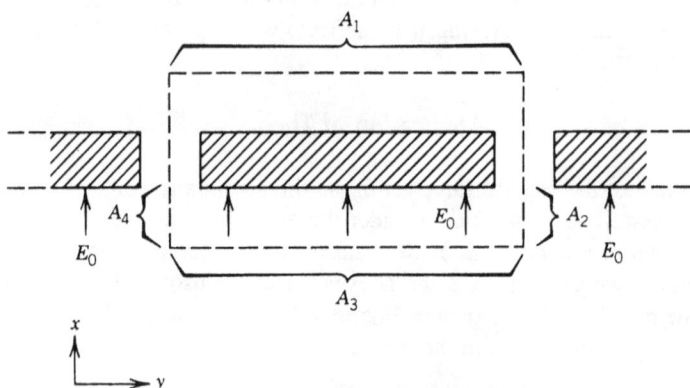

FIGURE 9.3.2. Electrode with embedded membrane.

gral over the second side may be written as

$$\iint_{A_2} T_{1j} n_j \, dA = \iint_{A_2} T_{12} n_2 \, dA = 0 \tag{9.3.27}$$

This integral vanishes because $T_{12} = \epsilon E_x E_y = 0$ since $E_y = 0$. On the third side, the normal vector is

$$n_x = -1, \qquad n_y = 0 \tag{9.3.28}$$

and the integral is

$$\iint_{A_3} T_{1j} n_j \, dA = \iint_{A_3} T_{11} n_1 \, dA = -\frac{\epsilon}{2} E_0^2 A_3 \tag{9.3.29}$$

On the fourth side the normal is

$$n_1 = 0, \qquad n_2 = -1 \tag{9.3.30}$$

and the contribution to the area integral is

$$\iint\limits_{A_4} T_{1j} n_j \, dA = \iint\limits_{A_4} T_{12} n_2 \, dA = 0 \qquad (9.3.31)$$

Adding all of the contributions gives the total force in the x direction as

$$f_x = -\frac{\epsilon}{2} E_0^2 A_3 \qquad (9.3.32)$$

The forces in the other directions can be computed in a similar fashion; the result of these computations is zero, so that the only force acting on the electrode is perpendicular to the electrode surface with a force per unit area (or stress) of

$$T = \frac{F_1}{A_3} = -\frac{\epsilon E_0^2}{2} \qquad (9.3.33)$$

Discussion

It should be noted that the stress on the membrane is *not* just the product of the charge on the membrane and the electric field as might be expected from the basic form of the electrostatic force equation. The charge on the surface is given by Gauss' law as

$$\rho_s = D = \epsilon E_0 \qquad (9.3.34)$$

and the product of the charge and electric field would therefore be twice as large as the actual force

$$\rho_s E = \epsilon E_0^2 \neq \tfrac{1}{2}\epsilon E_0^2 \qquad (9.3.35)$$

In general, the basic form of the force law $F = \rho E$ can be used only under the conditions for which it was derived, namely, a vanishingly small volume of charge in an externally generated field. In this device part of the field is set up by the charges on the membrane, so the force must be calculated by a method such as the stress tensor, which is valid for any charge distribution.

A second point should be emphasized in this result. The stress tensor integration gives the net force acting on all of the charges contained within the closed surface. It says nothing, however, about the force acting on the membrane. To relate the electrostatic force to the force on the membrane, we must make some assumption about the relation of the charges to the membrane. Here we assumed that the charges could not leave the membrane, so that all of the force acting on the charges was transmitted directly to the membrane. If the charges were free to leave, then the force would be applied to increase the momentum of the charges, or to overcome the drag of the surrounding medium, or perhaps to set the medium itself into motion. Since the actual relation between the charges and the material with which they interact can take on many different forms, the interaction assumption should be explicitly stated whenever the Maxwell stress tensor is used.

BIBLIOGRAPHY

Davies, M., Ed., *Dielectric and Related Molecular Processes*, Vols. 1–3, Chemical Society, London, 1972.

Holland, R., and E. P. Eer Nisse, *Design of Resonant Piezoelectric Devices*, MIT Press, Cambridge, MA, 1969.

Jones, T. B., Dielectrophoretic force calculation, *J. Electrostat.*, **6**: 69–82 (1979).

Lang, S. B., *Sourcebook of Pyroelectricity*, Gordon and Breach, New York, 1974.

Lang, J. L., and D. H. Staelin, Electrostatically figured reflecting membrane antennas for satellites, *IEEE Trans. Auto. Control*, **AC-27**: 666–670 (1982).

Melcher, J. R., *Continuum Electromechanics*, MIT Press, Cambridge, MA, 1981, Chap. 3.

Melcher, J. R., and M. Hurwitz, Gradient stabilization of electrohydrodynamically oriented liquids, *J. Spacecraft Rockets*, **4**: 864–881 (1967).

Nyberg, B. R., et al., Measuring electric field by using pressure sensitive elements, *IEEE Trans. Elec. Insul.*, **EI-14**: 250–255 (1979).

Pickard, W. F., Electrical force effects in dielectric liquids, *Prog. Dielectr.*, **6**: 1–39 (1965).

Pohl, H. A., *Dielectrophoresis*, Cambridge University Press, Cambridge, 1978.

Pohl, H. A., and K. Pollock, Electrode geometries for various dielectrophoretic force laws, *J. Electrostat.*, **5**: 337–342 (1978).

Sharbaugh, A. H., and G. W. Walker, The design and evaluation of an ion-drag dielectric pump to enhance cooling in a small oil-filled transformer, *IEEE-IAS Conf. Rec.*, Mexico City, October 1983, pp. 1161–1165.

Stuetzer, O. M., Ion drug pumps, *J. Appl. Phys.*, **31**: 136–146 (1960).

Woodson, H. H., and J. R. Melcher, *Electromechanical Dynamics*, Vol. 2, Wiley, New York, 1968, Chap. 8.

PROBLEMS

PROBLEM 1 (COMMUNICATIONS)

An electret loudspeaker is made from a layer of plastic of thickness a (= 30 μm) and dielectric constant κ (= 3) backed with a metal foil and stretched in front of a rigid electrode with the plastic facing the electrode at a distance b (= 60 μm). Find the electric stress applied to the foil/plastic laminate. You may neglect the motion of the laminate, which maintains a surface charge of ρ_s (= 20 μC/m^2).

PROBLEM 2 (AEROSPACE)

Electrostatic forces are now used in sophisticated optical devices to provide real time control and processing of optical and radio signals. For example, a diffraction grating can be synthesized by stretching a reflecting, conducting film over an electrode structure as shown in Figure 9.P.2. The electrode applies a voltage which varies across the width of the membrane so that the electrostatic force also varies.

The separation d is much smaller than λ, the wavelength of the applied voltage, so that the electric field has only a vertical component. If the membrane is under a

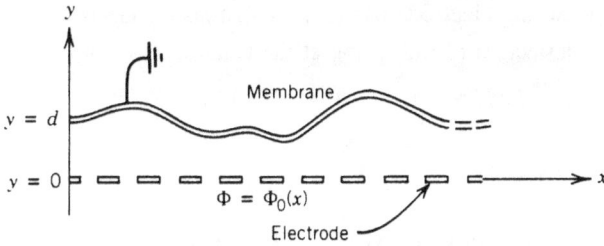

FIGURE 9.P.2. Electrostatically configured membrane.

tension K, its height $y(x)$ is given by

$$K\frac{\partial^2 y}{\partial x^2} = -\frac{f}{\text{area}}$$

Find $y(x)$ if $\Phi_0(x) = V \sin(2\pi x/\lambda)$, using the stress tensor to evaluate the pressure difference across the membrane. Assume $d \approx y$ to linearize.

PROBLEM 3 (INDUSTRIAL)

Find the force exerted by the ion drag pump of Section 9.1 by using the stress tensor instead of the force density.

PROBLEM 4 (INSTRUMENTATION)

A guarded electrostatic voltmeter is constructed as shown in Figure 9.P.4. An applied voltage tends to pull the center plate up, and the opposite end of the spring is pulled down until the electrode is again level with the guard electrodes.

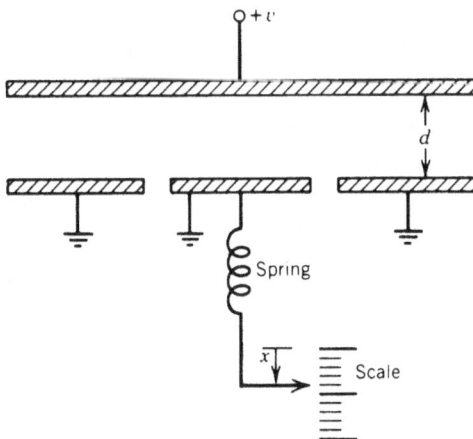

FIGURE 9.P.4. An electrostatic voltmeter.

a. If the measuring electrode has an area A, find the electric force on it.
b. Find the deflection of the spring at the balance if

$$f_{spring} = Kx, \qquad A = 10^{-2} \text{ m}^2, \qquad v = 10 \text{ kV}$$

$$K = 1 \text{ N/m}, \qquad d = 10 \text{ mm}$$

PROBLEM 5   (INDUSTRIAL)

Small and delicate metal pieces may be held for machining by electrostatic forces. A rough sketch of the contact area is shown in Figure 9.P.5a and modelled in Figure 9.P.5b.

a. What is the electric field in the air gaps?
b. What is the electric stress on the upper metal?
c. What is the leakage current i for a workpiece with an area of 10^{-4} m^2?

$$l = 10 \text{ }\mu\text{m}, \qquad b = 1 \text{ cm}$$

$$a = 0.1 \text{ }\mu\text{m}, \qquad \sigma = 10^2 \text{ S/m}$$

$$d = 5 \text{ }\mu\text{m}$$

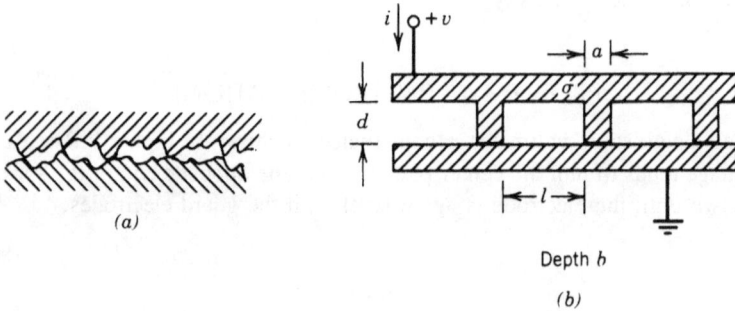

(a)

(b)

Depth b

FIGURE 9.P.5.   Electrostatic workpiece holder.

IV

ELECTROSTATICS OF CIRCUIT ELEMENTS

CHAPTER

10

RESISTIVE
CIRCUIT ELEMENTS

The uniform ohmic resistor is so straightforward that many of the more interesting applications of resistive elements are often overlooked. When the resistive material is nonlinear, for example, the device made from it will also be nonlinear, but the nature of the nonlinearity is substantially altered by the shape of the device. In fact, changes in the geometry of the device are often used to control the current flow in resistors. These changes may originate in subtle ways which are not obvious from the outside, but their analysis is not difficult once the basic mechanisms have been identified.

The resistive material is also subject to a number of physical and chemical effects which often form the basis for transducers in instrumentation. In many cases these same effects are influenced by the voltage and current through the resistive element so that a feedback interaction develops. These interactions limit the application of the transducers, but they also open the door for new applications of the resistors.

In this chapter the nonlinear interaction between material properties and geometry is illustrated by a varistor surge arrestor which is slightly deformed, resulting in poor performance. Variable geometry is often treated by a quasi-one-dimensional model of current flow. The most common device based on this interaction is the field effect transistor. The terminal relations of the FET are derived here using this model. Finally, the interactions among resistive terminal relations are illustrated by the thermal runaway which can occur in a thermistor, a resistive device often used to measure temperature.

10.1   NONLINEAR RESISTORS
(Varistor Surge Arrestors)

Summary

Many electrostatic devices employ electrically nonlinear materials. The terminal relations of these devices are naturally nonlinear, but not necessarily in the same way as the material. The geometry of the device interacts strongly with the material nonlinearity to produce different types of behavior at the terminals. This is illustrated by a bent surge arrestor.

Theory

Whether a resistor is made from ohmic materials or materials with nonlinear current relations, the terminal relations are determined in the same way. For the general device shown in Figure 10.1.1, the terminal relation is based on the conservation of charge. For a two-terminal device, the terminal current flowing into the positive terminal is given by

$$i = \iint \mathbf{J} \cdot dA \tag{10.1.1}$$

For a resistor, the processes are supposed to be occurring at a rate slow enough so that capacitive current is negligible. If this is not so, the methods discussed in Chapter 11 will have to be used.

If the geometry is fixed, the only additional information needed is the relation between the current density $\mathbf{J}$ and the applied voltage. In general the resistive material satisfies a constitutive equation of the form

$$\mathbf{J} = \mathbf{J}(\mathbf{E}) \tag{10.1.2}$$

The current density needed for the integral of Eq. (10.1.1) must be evaluated in terms of the electric field at the surface of the electrode. This field is obtained

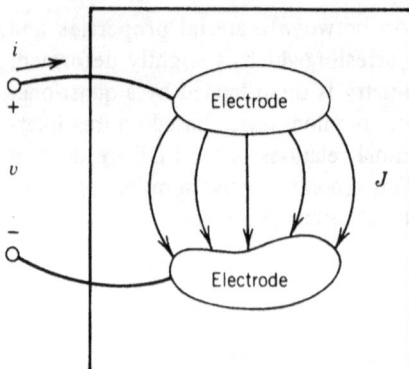

FIGURE 10.1.1.   A general resistive element.

from a solution of the electrostatic field equations, following the techniques discussed in the first part of this book. At the end of the procedure outlined here, the terminal relation will be given in the form

$$i = i(v) \tag{10.1.3}$$

It is important to realize that this terminal relation will not, in general, have the same form as the material relation for resistive material, Eq. (10.1.2). Whenever the material is nonlinear, there will be additional distortions introduced by the geometry itself unless the device is constructed with a simple parallel plate geometry.

Example: Varistor Surge Arrestor

The need for fuses or circuit breakers to protect circuits from excessive current flow is common knowledge. In many circuits, ranging from power transmission lines to personal computers, protection from excessive voltages is also required, since the insulators may be damaged above a critical voltage level even though the total current into the device is within the normal operating range.

An overvoltage protection device is, in a sense, the opposite of a fuse. When its voltage limit is exceeded, it must reduce its resistance to allow current to flow, rather than shutting the current off, as a fuse would. Ideally, an overvoltage device becomes a short circuit when it operates, allowing current to flow around the element to be protected. At low voltage levels, Zener diodes often serve as overvoltage protectors, whereas varistors (voltage controlled variable resistors) can perform the same function at the high voltage levels encountered in electric power transmission.

An idealized constitutive relation for ZnO, a material commonly used for varistors, is

$$J = 0, \qquad\qquad E < E_{cr} \tag{10.1.4a}$$

$$J = \sigma(E - E_{cr}), \qquad E > E_{cr} \tag{10.1.4b}$$

Below the critical field, very little current flows, and the material behaves as a very good insulator. If the electric field exceeds E_{cr}, however, the current increases quickly, limited only by the internal resistance of the material. This type of terminal relation is fairly common in devices which are subject to electric breakdown, such as corona discharges, semiconductor diodes, and gas discharge tubes. To be effective as overvoltage protectors, the onset of conduction should occur abruptly as the rated voltage is reached. This requires careful control of the geometry, as shown in the following discussion.

Application of the Theory

Varistor overvoltage protectors are usually constructed as a stack of wafers long enough to prevent air breakdown or surface tracking along the sides of the device.

If the device is a solid cylinder of length a and area A, the electric field inside is uniform with a magnitude of

$$E = \frac{v}{a} \tag{10.1.5}$$

and the net current to the positive electrode is obtained by integrating the current density, given by the piecewise linear relation of Eq. (10.1.4). Since the electric field is uniform, the current density at the electrode is also uniform, and the total current will have the same form as the current density,

$$i = 0, \qquad\qquad v < E_{cr}a \tag{10.1.6a}$$

$$i = \sigma A\left(\frac{v}{a} - E_{cr}\right), \qquad v > E_{cr}a \tag{10.1.6b}$$

This terminal relation is shown in Figure 10.1.2 by the curve labeled straight.

Because the device is so long in proportion to its width, it is possible that it can bend, which would destroy the uniformity of the electric field inside. A simplified model of this bending involves two parallel varistors, one shorter than the original and one longer, as shown in Figure 10.1.3. Because the section on the left is shorter, the electric field inside it

$$E_1 = \frac{v}{(a - \delta)} \tag{10.1.7}$$

is always larger than the electric field in the right section

$$E_2 = \frac{v}{(a + \delta)} \tag{10.1.8}$$

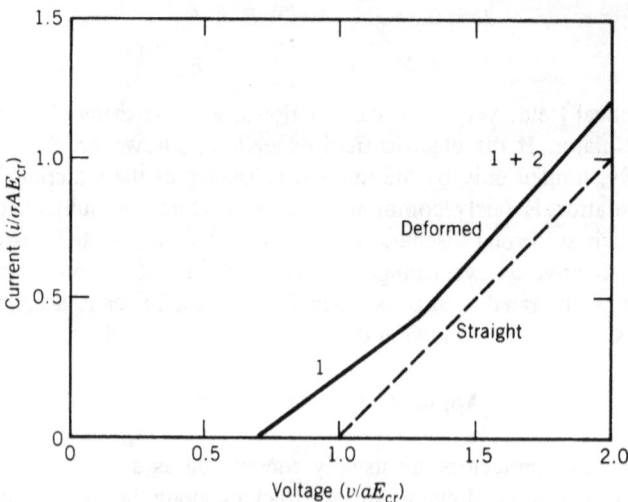

FIGURE 10.1.2. Terminal relation for nonuniform varistor.

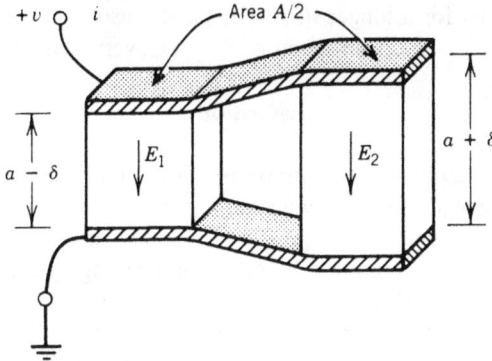

FIGURE 10.1.3. Model of a nonuniform varistor.

As the terminal voltage is raised, the section on the left will reach and exceed the critical field first, so it will start to conduct current while the other one is still blocking. Because only half of the total device is conducting, the current flow will be less, but it is still significant. The current flow attributed to this branch of the device i_1 is

$$i_1 = 0, \qquad\qquad v < (a - \delta)E_{cr}$$

$$i_1 = \frac{\sigma A}{2}\left(\frac{v}{a - \delta} - E_{cr}\right), \qquad v > (a - \delta)E_{cr} \qquad (10.1.9)$$

This current contribution is shown by the curve marked 1 in Figure 10.1.2.

As the voltage is raised still more, the second section eventually reaches its breakdown field, and it begins to conduct, adding its current to that of the first branch. Because this section is longer, it has a higher resistance to current flow, so the current does not increase as rapidly with voltage. The total current for the varistor includes both of these contributions, and is given by

$$i = 0, \qquad\qquad 0 < v < (a - \delta)E_{cr}$$

$$i = \frac{\sigma A}{2}\left(\frac{v}{a - \delta} - E_{cr}\right), \qquad (a - \delta)E_{cr} < v < (a + \delta)E_{cr}$$

$$i = \sigma A\left(\frac{a}{a^2 - \delta^2}v - E_{cr}\right), \qquad (a + \delta)E_{cr} < v < \infty \qquad (10.1.10)$$

This entire terminal relation for the deformed varistor is also shown in Figure 10.1.2 by the curve labelled deformed.

Comparing the total current for the nonuniform device with that for the cylindrical device shows that the current starts to flow at a lower voltage, but the device does not reach its maximum conductance until a much higher voltage. Both of these effects degrade the performance of the device. The leakage current at lower voltages causes ohmic heating of the device, which leads to deterioration, whereas the higher voltage needed to get full current flow means that overvoltage conditions will persist for a longer time. For these reasons it is important to avoid any nonuniformities in the construction of varistor overvoltage protectors.

Discussion

The change in the shape of the terminal relation with change in geometry is a common occurrence when dealing with nonlinear materials. As such, it is just a manifestation of the failure of superposition techniques in nonlinear systems. This has always been a major stumbling block in engineering applications of electrostatics, partly because nonlinear devices are so common in this area and partly because many people working in the area received most of their training in terms of linear circuits or systems. When confronted with a new device, like a varistor, the natural reaction is to go back to the techniques which have worked so well in describing similar but simpler devices such as resistors. Unfortunately, a great deal of the intuition developed under simpler circumstances leads to incorrect generalizations, so that nonlinear devices must always be approached with some wariness.

The basic effect of nonuniformities in a nonlinear material has been presented here for the simplest possible geometry, in keeping with the emphasis in this book on physical, rather than mathematical effects. The method can easily be extended to more complicated geometries, however, so long as the electric fields at the electrodes can be related to the terminal voltage. In practice, this usually requires analytical or numerical solutions of the electrostatic field equations in the actual geometry, followed by integration over the electrode surface to get the current.

10.2   RESISTORS WITH VARIABLE GEOMETRY
(Field Effect Transistors)

Summary

Many lumped resistors have an internal geometry which can vary during operation. The modeling of such devices is usually more difficult because the internal electric fields change with the geometry. A common approach to these geometrical effects is the quasi-one-dimensional model, which retains the simplicity of uniform field geometry with variable dimensions. This technique is used to derive the terminal relations for a junction field effect transistor (FET).

Theory

Even ohmic circuit elements can be affected by changes in their geometry. Simple examples of this effect, such as rheostats, have been known for a long time, but there are also newer, more complex devices which find wide application. The approach to modeling such devices follows the same pattern described in the previous section. The material constitutive equation relates the current density to the electric field, which is obtained by solutions of the electrostatic field equations with the terminal voltage as the boundary condition. Once the current density is known, it is integrated over an electrode to find the terminal current in terms of the applied voltage.

When the principal concern is the change in geometry, the simple uniform field solutions used before are rarely applicable. Instead, a two- or three-dimensional geometry which changes during operation must be used to find the field. In general, this leads to terminal relations which become quite complicated, and therefore useless for design. In many devices, however, a simpler geometrical model which includes variability but avoids the complications of two- or three-dimensional solutions of Laplace's equation is possible. This approach, which is sometimes called a quasi-one-dimensional model, is examined in the following discussion.

The key assumption here is that the current flow is mostly in one direction, although the conductor shape may vary in the current direction, as shown in Figure 10.2.1. Concentrating on the small slice of the conductor at the position x, we can write the approximate equations

$$E(x) = -\frac{d\phi}{dx} \tag{10.2.1}$$

$$i = J(x)A(x) \tag{10.2.2}$$

In the last equation, the total current passing any point $J(x)$ is independent of x because current must be conserved in the steady-state operation of a resistive element. The two equations for current and voltage are related by an ohmic constitutive relation

$$J(x) = \sigma E(x) \tag{10.2.3}$$

Combining and rearranging the equations gives

$$-\frac{i}{A(x)} = \sigma \frac{d\phi}{dx} \tag{10.2.4}$$

which can be integrated from one end of the device to the other to give the terminal relation.

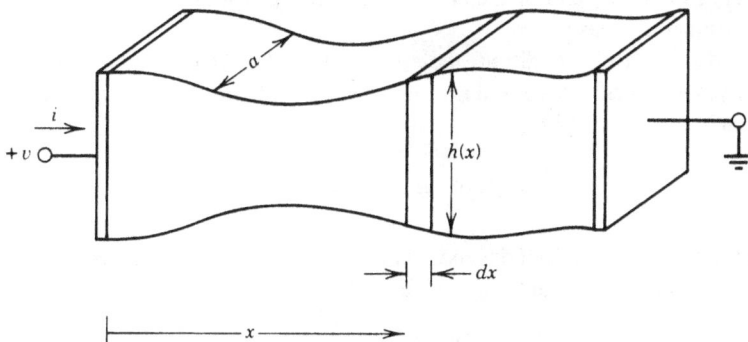

FIGURE 10.2.1.   Quasi-one-dimensional current flow.

Example: The junction FET

The junction field effect transistor (FET) is a device which depends on geometry changes to control the current flowing between two terminals, called the source and the drain (Fig. 10.2.2). In this device the channel which carries the current is bordered by a *p-n* junction, which is normally reverse biased. Under these conditions the channel adjacent to the junction is depleted of charge carriers and acts like an insulator, forcing the current to flow around it through the restricted region of width $d - a(x)$, which is below it in the figure. The size of this depletion layer depends on the amount of reverse bias, so the voltage applied to the gate electrode can be used to restrict or cut off current flow from source to drain. Since the *p-n* junction is reverse biased, this electrode, called the gate, requires very little power to operate, although it controls a large current flow.

The width of the depletion region is given in terms of the junction parameters. A typical relation might take the form

$$a(x) = \left[\frac{2\epsilon V_j}{qN_D}\right]^{1/2} \tag{10.2.5}$$

where N_D is the density of donor atoms in the semiconductor. The junction voltage V_j is measured from the gate electrode to the conducting channel. Since the channel is carrying current parallel to the junction, there is a voltage gradient along the channel, and the junction voltage will therefore vary along the junction,

$$V_j = \Phi(x) - V_g \tag{10.2.6}$$

The depletion layer will therefore vary in thickness, being widest near the drain where the reverse bias is largest.

The depletion layer will also widen as the gate voltage increases. In fact, at the critical voltage

$$V_d - V_g = V_p = \frac{qN_dD^2}{2\epsilon} \tag{10.2.7}$$

called the pinch-off voltage, the depletion layer grows to block entire width of the channel. Because of the voltage gradient along the channel, the pinch-off condition occurs first at one end of the channel and spreads toward the other end as the junction voltage increases.

Below the pinch-off condition, the current flows through the quasi-one-dimensional channel whose sides are deformed by the depletion layer. The area of the channel at any point is given by

$$A(x) = [d - a(x)]b = bd\left[1 - \left(\frac{\Phi(x) - V_g}{V_p}\right)^{1/2}\right] \tag{10.2.8}$$

Since the cross-sectional area depends on the channel voltage at that point, the differential form of the terminal relation, Eq. (10.2.4) takes the form

$$-\left[1 - \left(\frac{\phi - V_g}{V_p}\right)^{1/2}\right]d\Phi = \frac{i\,dx}{bd} \tag{10.2.9}$$

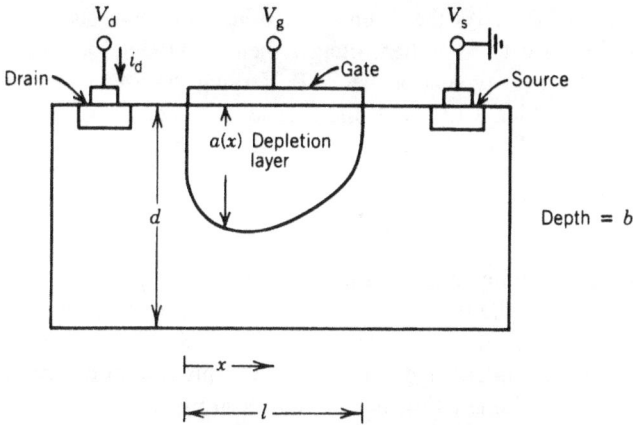

FIGURE 10.2.2.   A junction FET.

which can be integrated from the source at $x = 0$ and $\Phi = 0$ to the drain at $x = l$ and $\Phi = V_d$ to give the terminal relation

$$i_d = \frac{\sigma bd}{l}\left[V_d + \frac{2V_p}{3} - \left(\frac{V_g}{V_p}\right)^{3/2} - \frac{2V_p}{3}\left(\frac{V_d - V_g}{V_p}\right)^{3/2}\right] \qquad (10.2.10)$$

which is depicted in Figure 10.2.3. Only the solid portions of the curves are represented by the equation just derived.

There are four voltage terms in the terminal relation. The first is a linear term which corresponds to an ohmic resistor. It is dominant at low voltages where the

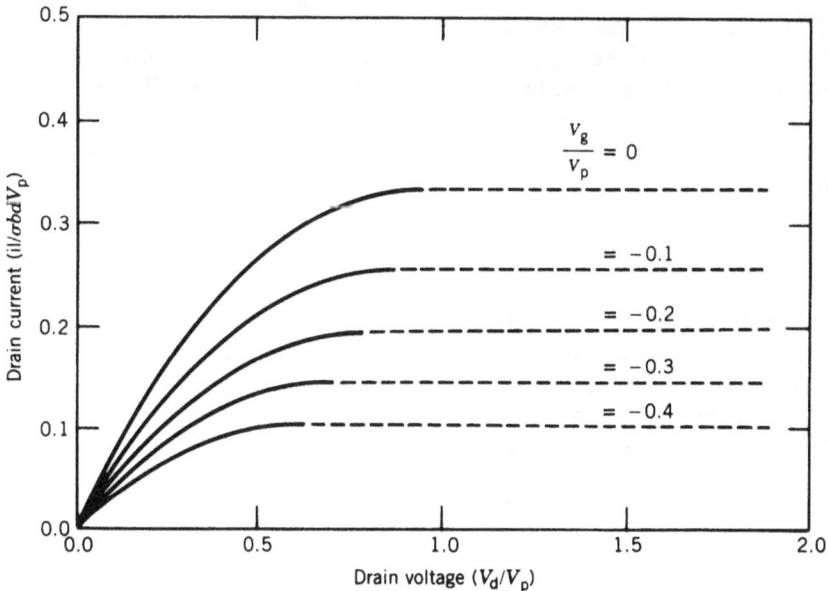

FIGURE 10.2.3.   FET terminal relations.

pinch-off effect vanishes. As the drain voltage increases, the depletion layer expands until it pinches off. At higher voltages the current increases very little, giving the horizontal curves typical of the FET. Varying the gate voltage will give a set of such curves, which saturate at different levels, depending on the additional channel voltage needed to reach the pinch-off condition.

Discussion

The FET is usually employed as an amplifier beyond the pinch-off region, where the curves are flat, and the device is well modeled as a voltage-controlled current source. It can also be used as a voltage-controlled resistor, since the slope of the curves depends on the voltage applied to the gate. This type of device is useful for such applications as volume adjustment under electronic control.

The quasi-one-dimensional method of handling nonuniform geometries is extremely useful in practice because it allows us to obtain terminal relations explicitly, a result which rarely follows from the direct application of two- or three-dimensional techniques. It also finds application in many other applications where the electrostatic fields control device behavior, such as capacitors and electron beams.

10.3   RESISTORS WITH VARIABLE MATERIAL PROPERTIES
(Thermistor Runaway)

Summary

Current conduction in resistive materials can be controlled by many more physical variables including temperature, magnetic fields, and humidity. Since these variables can in turn be affected by the current flow, there is a coupling among the parameters which is not always evident from the basic terminal relation. These interactions often lead to instabilities, such as the thermal runaway which occurs in thermistors.

Theory

The current flow through a resistive device can also change as a result of changes in the material constitutive relation brought on by changes in other physical parameters, such as temperature, pressure, humidity, and magnetic fields. For example, if temperature is the controlling variable, the terminal relation for a parallel plate geometry will have the form

$$v = v(i, T) \tag{10.3.1}$$

or, if the device is electrically linear,

$$v = R(T)i \tag{10.3.2}$$

Because the electrical variables are controlled by the temperature, a device of this type is a natural choice for a temperature transducer. Other transducers are also available for other types of controlling variables.

Ideally, these transducers should require only an ohmmeter to function, since the resistance is uniquely related to the input variable. In practice, however, some care must be taken in applications because the application of electric signals to the device may lead to alteration of the operating conditions. To obtain correct results, all of the interacting effects must be considered simultaneously.

Example: Thermistor Runaway

The thermistor is a resistor specifically designed to have an ohmic resistance which is a very strong function of temperature. It is widely used as a temperature transducer since it is more rugged than a thermocouple and operates at much higher signal levels. It is also used in precision circuits which must maintain a constant resistance over a wide range of temperature, since its decrease in resistance with temperature will offset the increase normally found in other conductors.

The negative temperature coefficient can cause problems, however, especially at large current levels. High currents heat the device and cause its resistance to fall. If a voltage source is applied, it is clear that the drop in resistance will lead to an increase in current, which will cause further heating. The process continues until some other factor, such as the external circuit or melting, finally limits the current. This is clearly unacceptable for a temperature transducer, so the ohmmeter must have some provision for limiting the current flow to a value which does not change the resistance noticeably.

Application of the Theory

The resistance of a thermistor is a nonlinear function of temperature since it is controlled by an activation energy. If stable operation is achieved, however, the temperature will not vary by large amounts, and the resistance can be represented by the linear approximation

$$R = R_0(1 - K_r T) \qquad (10.3.3)$$

The quantity T represents the rise in temperature above the ambient level. This rise depends on the input power to the device, $p = vi$, and, if thermal conduction is the dominant cooling mechanism, the temperature rise will be given by the linear relation

$$T = T(vi) = K_t vi \qquad (10.3.4)$$

The constant K_t will depend on the details of the heat conduction and generation and will vary from one situation to another.

Combining the thermal and electrical relations gives the terminal relation for the thermistor as

$$v = R_0(1 - K_t K_r vi)i \qquad (10.3.5)$$

or, solving for v,

$$v = \frac{R_0 i}{1 + K_t K_r R_0 i^2} \qquad (10.3.6)$$

which is plotted in Figure 10.3.1. At low voltages the device is linear, with the current increasing with voltage. At higher voltage levels the higher current heats the thermistor, reducing the resistance, which leads to even higher currents. Because of this there is a critical voltage which can not be exceeded in the device. It is given by differentiating Eq. (10.3.5) as

$$V_{cr} = \sqrt{R_0/(4K_t K_r)} \qquad (10.3.7)$$

A typical voltage, using the parameters

$$K_r = 0.01 \text{ K}^{-1}, \qquad K_t = 1500 \text{ K/W}, \qquad R_0 = 2000 \text{ } \Omega$$

is $V_{cr} = 5.8$ V, which occurs at a current of $I_{cr} = 5.8$ mA with a temperature increase of 50°C.

Discussion

When used as a temperature transducer, the thermistor should be kept far below the critical voltage and current. In other applications the nonlinear effects produced by heating may be desirable, and much larger currents are used. In these cases the device is acting like a switch, since it can have two values of current for a given voltage, depending on the external circuit and the way in which the voltage lever was reached. Of course, the external circuit should also be able to restrain the current from reaching damaging levels.

In some cases, devices with negative temperature coefficients do not have such protection, leading to a type of electrical breakdown called thermal runaway. Thermistors are subject to this instability, of course, but so are many relatively in-

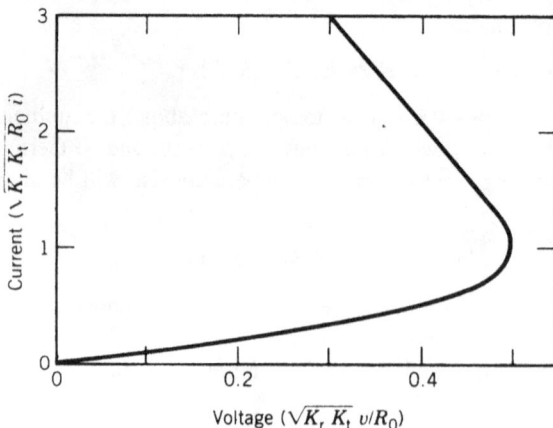

FIGURE 10.3.1. Terminal relation for a thermistor.

sulating devices in which leakage current is carried by ions or electrons excited from deep energy levels by thermal energy. Most insulators and many semiconductors exhibit this type of conductivity, and all of them will develop thermal runaway under the proper conditions.

BIBLIOGRAPHY

Gray, P. E., and C. L. Searle, *Electronic Principles*, Wiley, New York, 1969, Chap. 4.

Shaw, M. P., and N. Yildrum, Thermal and electrothermal instabilities in semiconductors, *Adv. Electr. Elec. Phys.*, **60**: 307–385 (1983).

Streetman, B. G., *Solid State Electronic Devices*, Prentice-Hall, Englewood Cliffs, NJ, 1972, pp. 285–301.

PROBLEMS

PROBLEM 1

A varistor is made of the same material described in Section 10.1 ($\sigma = 10^{-6}$ S/m, $E_{cr} = 1$ MV/m), which is mounted between two electrodes extending along radial lines from $r = a$ (= 1 cm) to $r = b$ (= 2 cm) with an angle θ_0 (= 0.5 rad) between them. The electrodes each have a depth d (= 1 cm). Find and plot the terminal relation for this device.

PROBLEM 2

Three light bulbs are connected in series across a 120-V source. A thermistor with the characteristics given in Section 10.3 is connected in parallel with each bulb. If one bulb burns out, the voltage across the thermistor rises above the critical voltage, which allows current to flow through the other two bulbs. What are the limits to the wattage ratings of the bulbs if they remain lit when the first one burns out?

CHAPTER
11

CAPACITIVE CIRCUIT ELEMENTS

Capacitors can become complicated devices in practice when their geometry and material properties change in operation. These changes may lead to nonlinear electrical behavior in which capacitance can be defined only with some ambiguity, depending on the nature of the applied voltage. The varactor diode, which can be used as a linear tuning element or as a nonlinear frequency multiplier, depending on the nature of the voltage, is an excellent example.

The difficulties are compounded by changes in geometry which generate new currents in addition to the currents associated with changing voltage. These currents increase with the speed of the motion, suggesting an alternate form of electromechanical energy conversion. Electrostatic generators using this principle have in fact been constructed for centuries and are still being designed for special applications, such as spacecraft. The dielectric material may also change in response to external influences such as pressure or temperature. These changes generate currents which can serve as a power source, as illustrated by a ferroelectric pulse generator.

The presence of multiple electrodes leads to further complications in capacitive devices since the charge on each electrode is influenced by the voltages on all of the other electrodes. Since the potential is conservative, any electrode potential can arbitrarily set equal to zero, a procedure which simplifies the problem by apparently eliminating one variable. A zero reference may not be appropriate for a particular problem, however, and, if it is made too early in the analysis, it may not be clear how to make a better choice. The procedure for choosing a good reference value is illustrated by a problem arising in ink jet printers.

11.1 NONLINEAR CAPACITORS
(Varactor Frequency Multipliers)

Summary

Nonlinear capacitive elements have more complicated terminal relations than do linear capacitors, but they also have more potential applications. Generally, these applications fall into two main classes. If the voltage variations are small, the device acts like an ordinary capacitance with a value controlled by the average level of voltage. If large voltage changes occur, the device behaves much differently, and the current may have little obvious resemblance to the voltage. This behavior is used to generate high-order harmonics in a varactor frequency multiplier.

Theory

The current that flows in a capacitor is commonly considered to be caused by changes in the voltage across the capacitor, which has the general terminal relation

$$q = q(v) \tag{11.1.1}$$

Thus the current is given in general by

$$i = \frac{dq}{dt} = \frac{\partial q}{\partial v}\frac{dv}{dt} \tag{11.1.2}$$

by the chain rule. This differential forms the basis of one common definition of the capacitance as

$$C_d = \frac{\partial q}{\partial v} \tag{11.1.3}$$

Unless the capacitor is linear, the differential capacitance will be a function of the voltage. If the voltage can be considered as a combination of a large bias voltage and a relatively small signal voltage of the form,

$$v = V_0 + v'(t) \tag{11.1.4}$$

The differential capacitance is a very useful concept. Expanding the current expression around the bias voltage gives

$$i = \frac{\partial q}{\partial v}\bigg|_{V_0}\frac{dv'}{dt} + \frac{\partial^2 q}{\partial v^2}\bigg|_{V_0}\frac{(v')^2}{2} + \cdots \simeq C_d(V_0)\frac{dv'}{dt} \tag{11.1.5}$$

If the signal voltage is small, the higher-order terms can be neglected, and the first-order term represents the flow of signal current. In this case the bias voltage V_0 is used to set the value of the differential capacitance, which is then used to tune or filter the signal voltage.

In general, however, this separation of voltage components is not allowed with a nonlinear device, and circuit analysis is more correctly handled by considering

the full nonlinear behavior. Since the current expression involves several derivatives, it is usually simpler to represent the device in terms of its terminal charge
rather than as a current. The circuit equations are then written as usual and solved
for the desired response.

Example: Varactor Frequency Multipliers

At microwave frequencies there is a shortage of good signal sources, so the high
frequencies needed must often be generated using approaches that avoid conventional oscillator circuits. One approach to this problem uses a nonlinear capacitor
called a varactor (for variable reactor). This is usually a junction diode which can
be reverse biased to give a large depletion layer whose thickness depends on the
applied voltage. When a voltage is applied, the depletion layer changes thickness
so that the charge storage changes nonlinearly with voltage. A typical terminal relation for such a device has the form

$$q = Cv\left(1 - \frac{v}{V_j}\right)^{-1/2}, \qquad v < V_j \tag{11.1.6}$$

which is plotted in Figure 11.1.1, along with an idealized version of the terminal
relation. As in any nonlinear circuit element, the current which flows has Fourier
harmonics at higher frequencies than those of the applied signal, and these harmonics can be selected to give an output signal at much higher frequencies than
the driving voltage. The current which flows in this device can be obtained by
solving the appropriate nonlinear circuit problem.

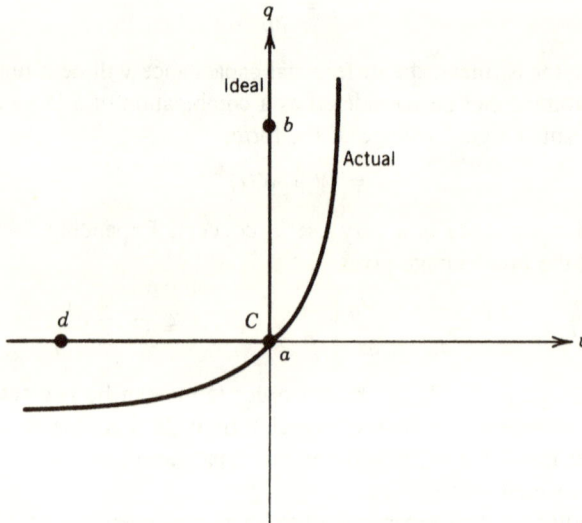

FIGURE 11.1.1.   Varactor terminal relations.

Application of the Theory

The basic circuit used to excite the high frequency harmonics is shown in Figure 11.1.2. The capacitor corresponds to the ideal nonlinear device whose terminal relation is given in Figure 11.1.1. The diode capacitor can not support a positive voltage and instead allows an infinite current flow, which would also correspond to an infinite amount of charge into the device, as represented by the vertical leg of the characteristic. When reverse biased, the diode is turned off, and the depletion layer becomes very wide, corresponding to a very small charge storage ability as represented by the horizontal leg of the characteristic.

The voltage source is usually a biased sine wave of the form shown in Figure 11.1.3, where the origin is placed at the point where the voltage first becomes positive. With positive voltage, the diode is on and presents no impedance to the current flow, which is limited only by the resistor. This current flow is given by

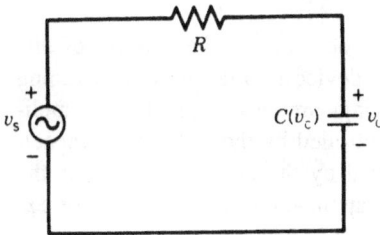

FIGURE 11.1.2. Varactor multiplier circuit.

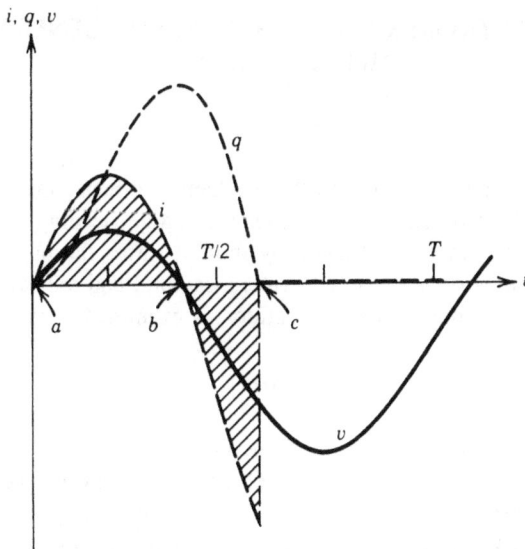

FIGURE 11.1.3. Voltage and current waveforms in a varactor multiplier.

$$i(t) = \frac{v(t)}{R}$$

and, as shown in Figure 11.1.3, it will have the same shape as the voltage as long as the diode is on. When the current is positive, charge flows into the diode starting at point a, in the figure. When the voltage becomes negative (point b), the current reverses, and charge flows out of the diode. This continues until all the charge injected into the diode (the area under the current curve a to b) is recovered.

When the charge level reaches zero, the diode operating point moves onto the horizontal leg of the characteristic, and the diode is turned off. It can no longer accept charge, so it acts as an open circuit, and the current drops abruptly to zero (point c). It is this abrupt current change that generates the high harmonics. Note that, although the input waveform is smooth and sinusoidal, the output current has a very sharp discontinuity due to the nonlinear nature of the varactor.

Discussion

As used here, the varactor is a highly nonlinear device, and the concept of differential capacitance is meaningless. The same device is often used as a tuning element in television sets, where the voltage swings are very small. In that application, the resonance of the tuning circuit is controlled by the differential capacitance of the varactor, which is in turn controlled by the bias voltage. Thus the same device can operate in a linear or nonlinear application, depending on the external circuitry.

11.2   ELECTROMECHANICAL CURRENT GENERATORS
(Rotating Generators)

Summary

The charge on a capacitor depends on the mechanical dimensions of the device. A change in any of these dimensions can alter the charge, leading to a current flow at the terminals. Current flow induced by motion forms the basis of electromechanical energy convertors such as generators, motors, and transducers. One such device, an electrostatic high voltage generator, is examined in more detail.

Theory

In the preceding section, current could flow only as a result of changes in voltage on the capacitor. In many applications such as transducers and generators, however, the spacing between the electrodes, their area of overlap, or some other aspect of the geometry changes, and these geometrical changes can all give rise to additional currents at the terminals.

When motion is allowed, the charge on the electrode will be expressed in the form

$$q = q(v, x) \tag{11.2.1}$$

The current associated with changes in this charge is, by the chain rule,

$$i = \frac{dq}{dt} = \frac{\partial q}{\partial v}\bigg|_x \frac{dv}{dt} + \frac{\partial q}{\partial x}\bigg|_v \frac{dx}{dt} \tag{11.2.2}$$

The first term is the usual capacitive current which flows when the voltage on the capacitor is changed, but it is supplemented by an additional current contribution which arises only when the geometrical variable x changes. This term is analogous to the speed voltage which occurs in electromagnetic generators in response to shaft rotation. Like the speed voltage, the speed current can also be used to generate electricity.

Example: A Rotating Electrostatic Generator

General Electric recently introduced an electrostatic high voltage generator intended for spacecraft applications (Philp, 1977). The basic arrangement of this generator can be illustrated by the simplified construction of Figure 11.2.1. This figure shows two electrodes separated by a distance d. One is fixed and one rotates on a shaft. When the angle θ is zero, the two electrodes line up to give the greatest capacitance at the terminals. As the rotor turns, the plates separate, and the capacitance decreases until the angle θ_0 is reached as shown in Figure 11.2.2. After this, the capacitance remains at zero until the electrodes again begin to overlap, and the capacitance increases toward its maximum value until one revolution has been completed.

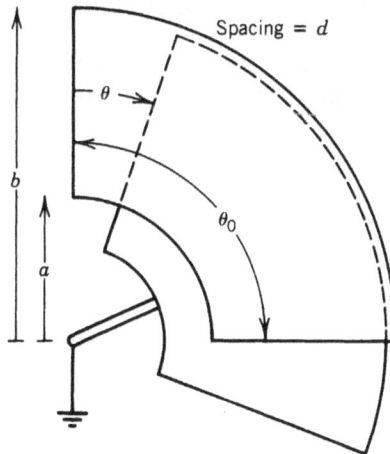

FIGURE 11.2.1.   A simplified electrostatic generator.

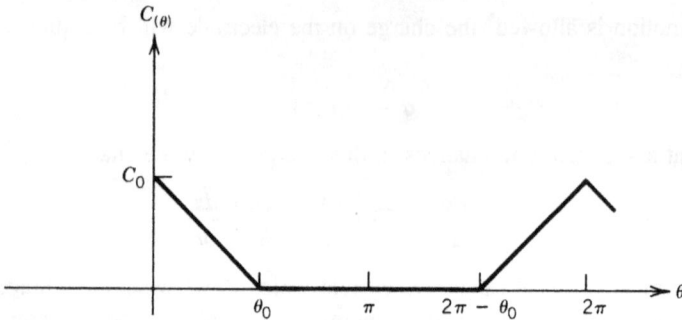

FIGURE 11.2.2. Angular dependence of capacitance.

In operation, the capacitor is connected to a voltage source which is sequenced so that net electrical power flows out of the terminals over one cycle of operation. Assuming that the rotor is turning at a steady rate corresponding to $\theta = \omega t$, one way of doing this is to short the terminals while the plates are approaching the maximum overlap and then to apply the full voltage V_0 at $\theta = 0$, charging up the capacitor. As the rotor continues to turn, the voltage is held constant while the plates separate. The voltage waveform corresponding to this mode of operation is shown in Figure 11.2.3. The primary question to be answered in such a device is the amount of electrical power generated in one cycle of operation.

Application of the Theory

The capacitance of the generator is a key factor in the conversion process. For the parallel plate geometry used here, the terminal relation is

$$q = \frac{\pi \epsilon_0}{d}(b^2 - a^2)\left(1 - \frac{\theta}{\theta_0}\right)v \equiv C_0\left(1 - \frac{\theta}{\theta_0}\right)v \qquad (11.2.3)$$

when $0 < \theta < \theta_c$. The terminal relation is similar for other segments, with an overall angular dependence shown in Figure 11.2.2.

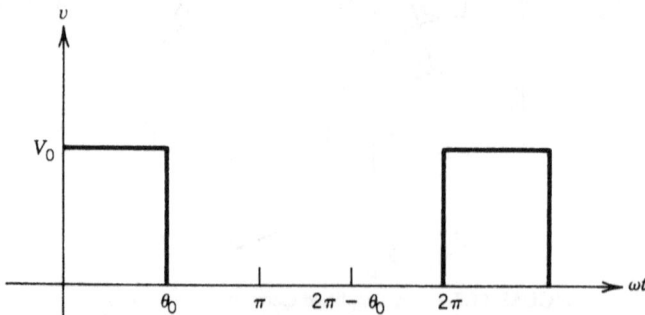

FIGURE 11.2.3. Terminal voltage on the generator.

The electrical power output at any instant is given by

$$P = -vi \tag{11.2.4}$$

The minus sign reflects the sign convention used here, in which current is assumed to flow into the positive terminal. Since

$$i = \frac{dq}{dt} \tag{11.2.5}$$

the total energy output over one cycle is given by

$$\frac{W}{\text{cycle}} = -\oint vi\, dt = -\oint v\, dq \tag{11.2.6}$$

To evaluate this integral, we need know only the voltage and current at every point in the cycle. Beginning shortly before $t = 0$, when the capacitance has its maximum value of C_0, the voltage is increased from zero to V_0. The generator does not have time to move significantly during this period, so the change in charge is given in terms of a fixed capacitance as

$$dq = C_0\, dv \tag{11.2.7}$$

Thus the contribution to the output of the generator during this brief initial period is $-\frac{1}{2}C_0 V_0^2$. The negative sign in the output indicates that electrical energy has been fed into the device.

Following the initial charging, the capacitor turns at constant voltage so that the change in charge is given by

$$dq = \frac{dq}{d\theta} d\theta = \frac{C_0}{\theta_0}(-d\theta)V_0 \tag{11.2.8}$$

Only the motion-induced current exists during this period since the voltage does not change. The contribution to the energy output is

$$\Delta W = -\int v\, dq = \int_0^{\theta_0} \frac{C_0 V_0^2}{\theta_0}\, d\theta = C_0 V_0^2 \tag{11.2.9}$$

For the remainder of the cycle the voltage is zero, so the power must vanish. Thus the net electrical energy output per cycle is

$$\frac{W}{\text{cycle}} = \frac{1}{2} C_0 V_0^2 \tag{11.2.10}$$

The output power is easily obtained by multiplying the net power per cycle by the number of cycles per second, which is given in terms of shaft speed as $\omega/2\pi$. As an example, consider a motor described in which

$$b = 2\text{ m} \gg a, \qquad d = 5\text{ mm}$$
$$V_0 = 100\text{ kV}, \qquad \omega/2\pi = 100\text{ s}^{-1}$$

The output power from this device would be approximately 600 kW if 100 of these sections are stacked in parallel on the shaft.

Discussion

Generation of currents by motion in electrostatic fields has a long history since it was originally the only practical way to convert mechanical energy to electrical energy. Benjamin Franklin developed such a machine, as did Robert van der Graaf, and even today electrostatic generation by motion is important, although mostly as a nuisance in such areas as electronics and petroleum transport, where the resulting high voltages can damage or destroy equipment.

Electrostatic generation also occurs in transducers, where the motion of charged electrodes can be translated into an electrical output signal. In most of these devices the generated current coexists with currents induced by changing voltages. Both of these currents must be considered, of course, when designing such devices.

11.3   CAPACITORS WITH VARIABLE DIELECTRICS
(Ferroelectric Pulse Generators)

Summary

In addition to changes in voltage and dimensions, changes in the electrical properties of the dielectric can also generate currents. These currents often seem mysterious since they are usually associated with anisotropic media, which are difficult to describe simply. They are really no more involved than the other currents, however, if the fundamental techniques are applied just as they were before on simpler materials. This approach is illustrated by a ferroelectric pulse generator.

Theory

In the preceding sections we have seen how changes in the voltage or dimensions in a capacitor can generate currents. Changes in the dielectric properties can also lead to currents at the terminal of the device. In devices where significant currents are generated by material changes, the terminal relation is likely to be very nonlinear, and it is usually essential to concentrate on the total charge rather than the terminal current to avoid the difficulties of dealing with currents which approximate impulses. In this section we also neglect, as far as possible, the currents generated by changes in voltage and geometry to concentrate on the new effects. In practice, of course, all of these effects may occur simultaneously.

We assume the basic parallel plate geometry for the capacitor with an electric field $E = v/a$ and a material relation

$$D = \epsilon_0 E + P(E) \qquad (11.3.1)$$

With nonlinear materials, there is little point in defining a permittivity, and the polarization is expressed separately as a function of the applied field. This leads to a terminal relation

$$q = \left[\frac{\epsilon_0 v}{a} + P\left(\frac{v}{a}\right) \right] A \qquad (11.3.2)$$

which, except for nonuniform geometric factors, is identical to the constitutive equation for the material. The polarization may depend on additional factors, so the terminal relation will often be a set of relations, as shown in Figure 11.3.1.

Either of the curves shown in this figure can be the terminal relation, depending on the particular circumstances. The charge varies with voltage, but in this case the charge also varies with some other parameter such as the pressure or the temperature. Because the device can present different terminal relations depending on some third parameter not present at the terminals, it is similar to many other controlled devices in electronics (e.g., transistors). Circuits involving these controlled capacitors can be solved by using techniques similar to those used in nonlinear transistor circuits.

There are some important differences between circuits with controlled capacitors and those with controlled resistive devices such as transistors, however. The terminal relation is expressed in terms of charge rather than current, so the usual Thevenin or Norton equivalent circuits can not be immediately used to specify load lines. Instead, the circuit will have both resistive and capacitive elements so that time variation (and the associated differential equations) become necessary much earlier in the solution.

Example: Ferroelectric Pulse Generators

Ferroelectric materials, like the ferromagnetic materials which lent them their name, can be permanently polarized in an external field. If a strong electric field is applied to such materials and then removed, the polarization remains indefinitely. This effect is usually explained in terms of ions in the material which can assume two stable positions, each slightly off center in the crystal, so that a dipole moment is produced. The external field pulls all of the ions into the position in the

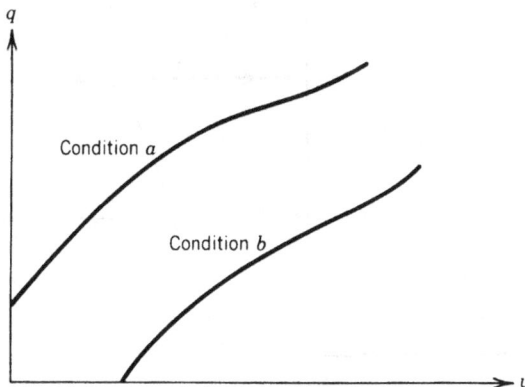

FIGURE 11.3.1.   A typical terminal relation for a controlled dielectric.

field direction, and they remain in this stable position when the field is removed (cf. Section 8.2). The terminal relation for such a device has the idealized form shown by curve *a* in Figure 11.3.2 when a positive electric field is applied. A strong field in the opposite direction can move all the ions to the other stable position so that the polarization reverses (along the line *b* in the figure). Continuing with a return to the original field direction, the device describes a hystersis loop.

Although the external field can align all of the dipoles if it is strong enough, the ions may be more stable if they alternate directions so that pairs of dipoles cancel each other out. In this situation, called the antiferroelectric state, the net polarization is very small. The terminal relation corresponding to this state is shown in Figure 11.3.2 as curve *c*.

A material such as lead zirconate titanate (PZT) can exist in either the ferroelectric or the antiferroelectric state over a range of voltages from zero to V_c, as shown in the figure. Experimentally, it has been found that a large mechanical pressure, such as a shock wave, can trigger a transition from the ferroelectric to antiferroelectric state, presumably by disturbing the individual dipoles enough so that they can rearrange themselves from the metastable parallel orientation to the more stable antiparallel orientation.

This effect has been used to store electric energy for long periods by first poling the ferroelectric in a strong field and then releasing the energy quickly by mechanical impact. For practical applications, the amount of energy stored and the nature of the output pulse are both important. The methods used to obtain these quantities are discussed in the following section.

Application of the Theory

This device will normally deliver its energy to a passive load, which is represented by a linear resistor in the circuit of Figure 11.3.3. This circuit is described by the

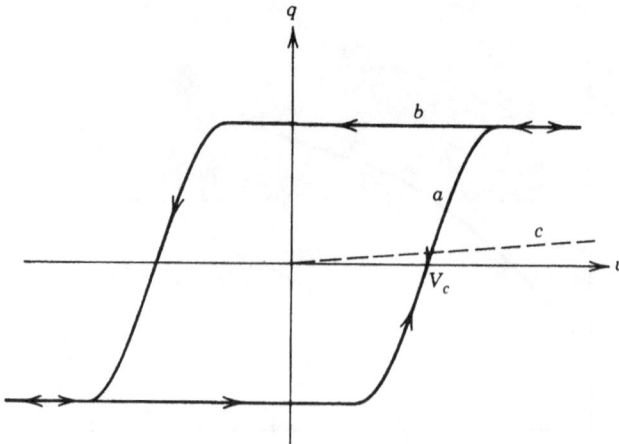

FIGURE 11.3.2. Terminal relations for an idealized ferroelectric device.

Ferroelectric
device

Pressure

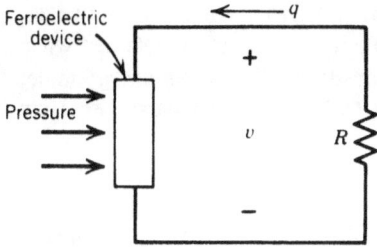

+

v R

−

FIGURE 11.3.3.   Ferroelectric pulse generator.

loop equation

$$v + R\frac{dq}{dt} = 0 \qquad (11.3.3)$$

The operating path of the generator is shown in Figure 11.3.4. Initially, no charge flows from the terminals, so the voltage is zero, and the device operates at point a in the figure. When the mechanical shock occurs, charge will flow out as the material is depolarized. The current through the resistor causes the voltage to rise above zero while the charge falls, following the trajectory labeled b until the transition is complete, and the voltage returns to zero at point c.

The electric power supplied to the load during the transition is

$$vi = v\frac{dq}{dt} \qquad (11.3.4)$$

and integrating this to find the total energy supplied gives

$$W = \int_{Q_p}^{0} v\,dq \qquad (11.3.5)$$

which is the shaded area in the figure. The output could be increased by enlarging this area, which would require the voltage to increase at any charge level during

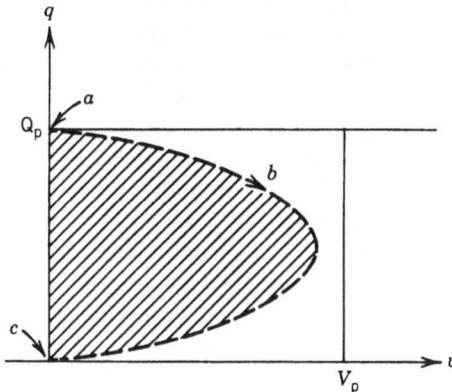

FIGURE 11.3.4.   Operating path of a ferroelectric pulse generator.

the transition. The voltage should not exceed the polarizing voltage V_p, however, or the transition would be halted. The maximum output is therefore obtained when the voltage is held just below V_p for the entire transition. This output energy under these conditions is $V_p Q_p$. For a typical material (PZT) made into a cube with 1-cm sides,

$$Q_p = 24 \ \mu C, \qquad V_p = 8.3 \ kV$$

so $W = 0.2$ J. This is sufficient energy to ignite an explosive mixture, which suggests such applications as emergency flares or fuses.

This maximum energy output will not be obtained in practice unless the operating trajectory is rectangular. The actual trajectory is obtained by solving Eq. (11.3.3), but this requires us to know the rate at which current is released in response to the mechanical shock. Lacking detailed information on this process (which is described more fully by Berlincourt, 1968), we could assume that the polarization is rearranged at a uniform rate over a depolarizing time t_d so that

$$i = \frac{dq}{dt} \simeq \frac{Q_p}{t_d} \qquad (11.3.6)$$

Constant discharge rate corresponds to a model in which the depolarization proceeds very quickly once the pressure wave arrives. Since the acoustic pulse travels at a constant rate, it will require a finite time to traverse the material. For this example, $t_d = 2 \ \mu s$ with a typical sound velocity of 5 km/s.

If the discharge rate is constant during the transition, the voltage across the resistor will remain constant, corresponding to the desired trajectory shape if the resistor is chosen so that

$$R = \frac{t_d V_p}{Q_p} \simeq 700 \ \Omega \qquad (11.3.7)$$

Discussion

In practice, the actual trajectory will not be as simple as the one assumed here, since it will depend on the details of the electromechanical interaction. The actual solution (Halpin, 1966) involves such details as the propagation of the mechanical wave through the solid, as well as additional circuit elements. These extensions of the model, however, will not change the basic conclusions concerning the amount of energy available and the best trajectory for obtaining it.

The example concentrates on a transition from ferroelectric to antiferroelectric states brought on by mechanical stress which disorders the aligned dipoles and allows them to settle into the more stable antiparallel orientation. Since the disordering can also be caused by thermal agitation, this device has been suggested for thermal energy conversion. At present, however, it does not appear to be efficient enough to be generally useful, although it may serve as an emergency, one-shot supply.

11.4   MULTIPLE TERMINALS AND MUTUAL CAPACITANCE
(Droplet Charging in Ink Jet Printers)

Summary

Capacitors with two terminals are relatively simple because there is only one potential difference to consider. With three terminals, there are three potential differences, and the number of choices rises quickly as more terminals are added. To limit the number of variables, the potential of the individual electrodes, rather than the potential difference, is usually selected as the electrical variable. This requires an additional choice of a reference potential, which must be made explicit to avoid confusion. The process is illustrated by the charging of drops in an ink jet printer.

Theory

So far, all of the lumped electrostatic devices have been represented by ordinary capacitors with just two terminals. In many practical devices, however, there are more than two terminals, which may be connected in various ways to external sources. The simplest case, which involves just three terminals, is considered here in some detail, since extensions to more than three terminals are relatively straightforward.

Consider a three-terminal electrostatic device, which is shown in general form in Figure 11.4.1. Each of the terminals is connected internally to a separate electrode. Conservation of charge still applies to each of these electrodes, so the rela-

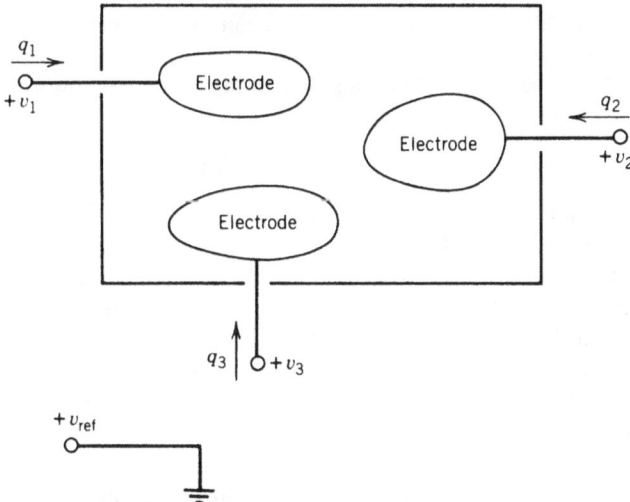

FIGURE 11.4.1.   A general three-terminal device.

tion which describes that terminal will always have the general form

$$i_i = \frac{dq_i}{dt} \tag{11.4.1}$$

The net charge on the electrode is obtained, as before, by Gauss' law over the surface of the ith electrode

$$q_i = \oint_{A_i} \mathbf{D} \cdot d\mathbf{A} \tag{11.4.2}$$

The only novelty in the procedure comes because the field at the surface of the ith electrode will depend, in general, on the voltages at each of the electrodes in the device,

$$\mathbf{D}_i = \mathbf{D}_i(v_1, v_2, v_3) \tag{11.4.3}$$

and the charges will likewise depend on all electrode voltages,

$$q_i = q_i(v_1, v_2, v_3) \tag{11.4.4}$$

Although all three voltages appear in this expression, we are not dealing with three independent quantities. The electric potential is a conservative function, and it is only the potential difference which is important in practice. Often the terminal relations are simplified by selecting the voltage reference to be zero at one of the terminals. For example, if the third electrode, on the bottom in the figure, is selected as the reference $v_3 = 0$ only the first two voltages will appear explicitly in the terminal relations, which then take the form

$$q_i = q_i(v_1, v_2) \tag{11.4.5}$$

This simplifies the description of the device at the price of some loss in generality, since it assumes a voltage reference which may not be appropriate in some applications.

In many devices the relation between charge and voltage is linear, so that the terminal relations take the form

$$q_1 = C_{11}v_1 + C_{12}v_2 + C_{13}v_3 \tag{11.4.6}$$

using the first terminal as an example. The coefficient of v_1 is called the self-capacitance C_{11} of the first electrode and corresponds to the capacitance measured when all of the other electrode voltages are held at zero. The second coefficient C_{12} is a mutual capacitance which involves charge induced on the first electrode by voltage applied to the second electrode, again when all other voltages are set to zero. In a manner analogous to mutual inductances, these mutual capacitances allow a transfer of charge from one terminal to another, an effect which must be considered when using multiple electrodes.

Example: Charging Ink Drops in an Ink Jet Printer

One application where mutual capacitance has played a crucial role is the ink jet printer. Ink drops are individually charged to control their trajectory toward their destination on the paper. This charge is applied to the drops by a concentric electrode as they break off from an ink jet. Since each drop must be deflected to a different point on the paper, successive drops usually receive different charges. As one drop breaks off, charge is induced by the charging electrode, but it is also induced by the preceding drop charge, which changes from one drop to another. Thus the individual drop charge may vary even though the same charging voltage is applied to the concentric electrode. This effect is large enough to cause serious errors in drop placement, with corresponding loss in image quality on the paper.

One way in which image quality can be restored is to remember (electronically) the charge given to the previous drop and to correct for it by altering the voltage on the charging electrode. This method is straightforward, and, since most of these printers are driven by microcomputers, it does not require a great increase in system complexity. It does require, however, that the disturbance induced by the previous charged drop be known in advance, and this can be determined from the mutual capacitances.

Application of the Theory

The geometry in an actual ink jet printer is fairly complex, involving the charging electrode, roughly spherical drops undergoing oscillations, and the disturbed jet connected by a thin ligament to the forming drop. To illustrate the method, we drastically simplify the geometry to the form shown in Figure 11.4.2 and neglect fringing fields. Just at the point of drop breakoff, there are four charge-carrying electrodes in this system. These are the drop which is just forming, identified by the subscript d, the previous drop p, the jet j, and the charging electrode e, which consists of an upper and lower part. The two drops are modeled as cubes, and the remaining electrodes are assumed to have the corresponding rectangular shapes shown in the figure.

The charge on the forming drop is the primary concern here. It is given by Gauss' law as

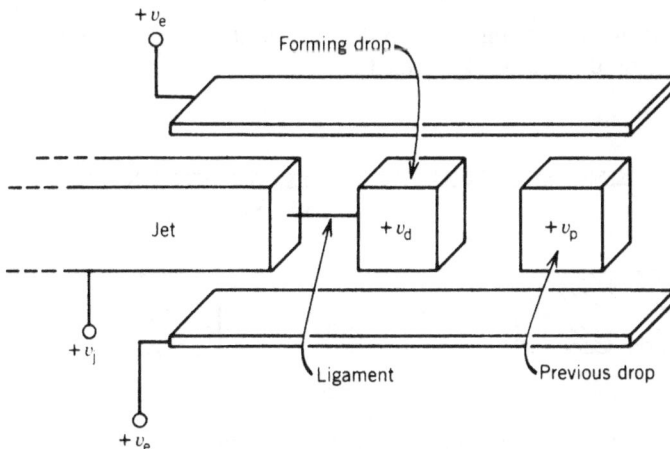

FIGURE 11.4.2. Simplified charging geometry for an ink drop.

$$q_d = \oint_{A_d} \mathbf{D} \cdot d\mathbf{A} \qquad (11.4.7)$$

The electric fields at the surface of this drop depend on the potentials of the neighboring electrodes as shown in Figure 11.4.3, where s is the drop separation and c is the distance to the charging electrode. These fields are directed as shown in the figure and have the magnitudes

$$E_c = \frac{v_d - v_e}{c} \qquad (11.4.8)$$

$$E_p = \frac{v_d - v_p}{s} \qquad (11.4.9)$$

$$E_j = \frac{v_d - v_j}{s} \qquad (11.4.10)$$

Using these fields in Gauss' law gives the net charge on the drop as

$$q_d = \frac{2\epsilon d^2(v_d - v_e)}{c} + \frac{\epsilon d^2(v_d - v_p)}{s} + \frac{\epsilon d^2(v_d - v_j)}{s} \qquad (11.4.11)$$

which has the form

$$q_d = C_{dd}v_d + C_{dp}v_p + C_{de}v_e + C_{dj}v_j \qquad (11.4.12)$$

Just at the point of breakoff, the jet and the new drop are connected by a conducting ligament, so they will be at the same voltage. Since we are allowed to pick one voltage level as a reference, the work that follows is simplified by assuming that the jet and the forming drop are both at the reference level (i.e., grounded)

$$v_j = v_d = 0 \qquad (11.4.13)$$

This simplifies the charge expression for the drop to

$$q_d = C_{de}v_e + C_{dp}v_p \qquad (11.4.14)$$

As anticipated, the charge on the drop depends on the charging voltage, but also on the voltage on the previously formed drop.

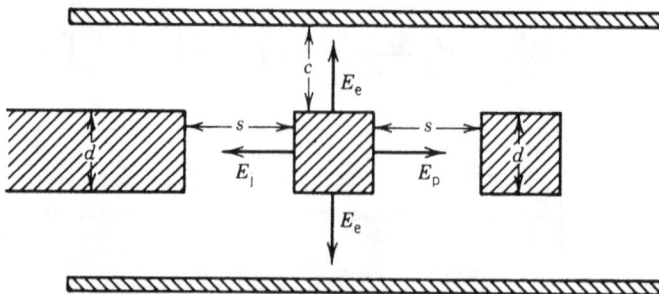

FIGURE 11.4.3. Electric fields around the charging drop.

This voltage v_p is not known in advance. Rather, the charge on the previous drop has been fixed, so we have to convert this charge into the corresponding voltage. This can be done with the terminal relation for the previous drop, which is found in the same manner by integrating Gauss' law over the drop surface. After integrating and using the same voltage constraints on jet and new drop, the charge on the previously formed drop is given by

$$q_p = \frac{2\epsilon d^2(v_p - v_e)}{c} + \frac{\epsilon d^2(v_p - v_d)}{s} \tag{11.4.15}$$

or, in terms of the capacitance coefficients and the known charge,

$$q_p = C_{pp}v_p + C_{pe}v_e \tag{11.4.16}$$

when $v_d = 0$ as before. Solving this equation for v_p and substituting into the charge expression for the new drop, Eq. (14.4.4), gives the charge on the new drop as

$$q_d = C_{de}\left(1 - \frac{C_{pe}C_{dp}}{C_{de}C_{pp}}\right)v_e + \frac{C_{dp}}{C_{pp}}q_p$$

$$= -\frac{4\epsilon d^2}{c}\left(\frac{s + c}{2s + c}\right)v_e - \frac{c}{2s + c}q_p \tag{11.4.17}$$

As expected, this charge depends on both the charging voltage and the charge on the previous drop. To charge the new drop accurately, the charging voltage must be controlled to account for the disturbance introduced by the previous drop.

Discussion

If microprocessor control of the charging voltage is too elaborate for the intended application, there are alternate ways of reducing this effect. The simplest is to separate the drops to be printed from each other by "guard drops" to which no charging voltage has been applied. The goal here is to ensure that the charge of the previous drop always has the same (low) value so that the charging corrections will be small enough to neglect. The guard drops will not be completely free of charge, of course, since their previous drop will have been given a specific deflection charge, but if the mutual capacitance between drops is small, the variation in charge may be negligible. In any case, the size of the error can be determined by using the same techniques described previously.

Although there are four electrodes in this problem, all of the charges of interest are given in terms of two voltages owing to the selection of a zero voltage level for the jet and the connected drop. This sort of simplification is sensible in a particular application since it reduces the algebra, but it should never be forgotten that it depends on particular terminal constraints. Once the drop separates from the jet, for instance, the simpler terminal relations are no longer valid, and the complete set must be used with the new terminal constraints. This is especially important later when we use the energy method for calculating the forces on these electrodes, since the terminal voltages must be unconstrained for this method to work.

BIBLIOGRAPHY

Angrist, S. W., *Direct Energy Conversion*, Allyn and Bacon, Boston, 1965, pp. 367–373.

Berlincourt, D. A., Piezoelectric and ferroelectric energy conversion, *IEEE Trans. Sonics Ultrason.*, **SU-15**: 89–97 (1968).

Bright, A. W., and B. Malsin, Modern electrostatic generators, *Contemp. Phys.*, **10**: 331–353 (1969).

Fillmore, G. L., et al., Drop charging and deflection in an electrostatic ink jet printer, *IBM J. Res. Dev.*, **21**: 37–47 (1977).

Jones, T. B., Scaled-up capacitive models for ink-jet printers, *J. Electrostat.*, **13**: 43–54 (1982).

Katz, H. W., *Solid State Magnetic and Dielectric Devices*, Wiley, New York, 1959.

Kettani, M. A., Ferroelectric power generation in *Direct Energy Conversion*, Addison-Wesley, Reading, MA, 1970, Chap. 11.

Mulcahy, M. J., and W. R. Bell, Electrostatic generators, in *Electrostatics and its Applications*, A. D. Moore, Ed., Wiley, New York, 1973, Chap. 8.

Philp, S. E., The vacuum-insulated, varying-capacitance machine, *IEEE Trans. Electr. Insul.*, **EI-12**: 130–136 (1977).

Streetman, B. G., *Solid State Electronic Devices*, Prentice-Hall, Englewood Cliffs, NJ, 1972, pp. 187–191.

Unger, H.-G., and W. Harth, Physics and applications of MIS varactors, *Adv. Electron. Electron Phys.*, **39**: 281–346 (1973).

Woodson, H. H., and J. R. Melcher, *Electromechanical Dynamics*, Vol. 3, Wiley, New York, 1968, Chap. 12.

PROBLEMS

PROBLEM 1 (ELECTRONICS)

The charge stored in the depletion layer of a *p-n* junction is given by an equation of the form

$$q = C_0 v \left(\frac{1 - v}{V_j} \right)^{-K}$$

where K depends on the structure of the junction. Find the differential capacitance as a function of bias voltage.

PROBLEM 2 (COMMUNICATIONS)

A capacitor microphone consists of two parallel electrodes of area A, initially separated by a distance d (= 1 mm). The electrodes are connected to a bias voltage V_0 (= 500 V) and a resistor R (= 100 MΩ) in series. Find the output voltage developed across the resistor as a function of frequency if the electrode separation varies sinusoidally with a peak displacement δx (= 10 μm).

PROBLEM 3   (POWER)

Electret generators are electric devices in which the rotor has a permanent charge and are similar in many respects to a permanent magnet motor. They can be modeled by the terminal relations

$$v_s = S_1 q_s + S_m q_r$$
$$v_r = S_m q_s + S_2 q_r$$

where

$$S_m = S_0 \cos \theta$$

(S is the elastance coefficient, which is the reciprocal of the capacitance.) In the steady state $q_r = Q_0$ and $\theta = \omega t + \gamma$.

   a.   What is the equivalent circuit of the stator?
   b.   What is the short circuit current?

PROBLEM 4

In the Kelvin water drop generator, two droplet streams develop a high voltage with the arrangement shown in Figure 11.P.4.

The buckets are made of plastic, with $\epsilon = 2.5\epsilon_0$. The area of the bottom is A ($= 0.1$ m^2), and the thickness is d ($= 1$ mm). The drops form from a grounded jet

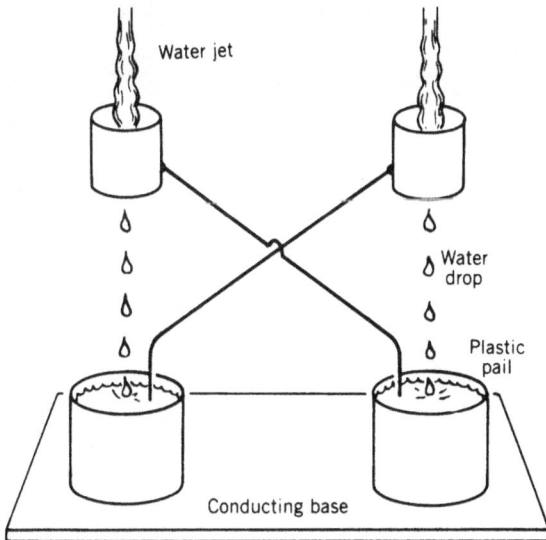

FIGURE 11.P.4.   Kelvin water drop generator.

of water of radius a (= 1 mm) while inside a charging cylinder of radius b (= 5 mm). The charging electrodes are connected to the water in the opposite buckets.

a. If buckets have charges Q_1 and Q_2, find the voltages v_1 and v_2.

b. If the droplets form from a section of the jet of length l (= 1 cm), find the charge on a droplet q_1 or q_2 in terms of v_1 or v_2.

c. If N (= 100) drops fall each second, find the voltages v_1 and v_2 as a function of time, assuming

$$v_1(t = 0) = 1 \text{ V}, \qquad v_2(t = 0) = 0 \text{ V}$$

CHAPTER

12

FORCES ON LUMPED
ELEMENTS

Electrostatic forces on charged particles can be calculated directly, and forces on charge distributions can be calculated with force densities, or stress tensors, but these methods all require a detailed knowledge of the fields within or on the surface of the region of interest. With a lumped element this information is often unavailable. In fact, only the terminal relations are usually given, so it would be far more useful to have a way to calculate the force from the terminal relations.

Fortunately, there is such a method, which is based on the conservation of energy. It can be used only for conservative devices, but most electrostatic devices which produce forces are conservative or can be treated as such by appropriate modifications of the model. If so, the energy method offers a quick method for calculating electrostatic force, requiring just one integral and one derivative.

The basic energy method can be further simplified by defining additional conserved quantities called coenergies. They allow more flexibility in performing the required integrals and are often used in devices with nonlinear dielectrics or multiple terminals.

The energy methods can be used for the linear forces which arise in devices like electrostatic speakers and for torques which occur in rotating devices like electrostatic display signs. Coenergy can also be used for linear forces or torques, which are illustrated by ferroelectric transducers and electrostatic motors, respectively.

12.1   LINEAR FORCES
(Electrostatic Loudspeakers)

Summary

The electrostatic force exerted by a conservative lumped circuit element is best calculated by an energy method rather than by direct integration of the force density. The energy method requires a carefully selected path of integration to avoid ambiguity but is fast and simple in practice. It is illustrated here by the calculation of the output of an electrostatic loudspeaker.

Theory

The energy method (Woodson and Melcher, 1968) is the most generally useful means of finding the force exerted by a lumped electrostatic device. It is based on the conservation of energy written for the device itself, neglecting any other electrical or mechanical attachments. Since this is a lumped device, energy can flow in or out only at specific terminals, as indicated in Figure 12.1.1. In this figure we assume a single electrical input, with electrical power flowing into the terminals on the left and mechanical power flowing out of the terminal at the right. The total energy stored inside the device changes as a result of these flows,

$$\frac{dW}{dt} = vi - f\frac{dx}{dt} \qquad (12.1.1)$$

In a capacitor the current is always given in terms of the stored charge as

$$i = \frac{dq}{dt} \qquad (12.1.2)$$

so power conservation, Eq. (12.1.1), can be written as

$$\frac{dW}{dt} = v\frac{dq}{dt} - f\frac{dx}{dt} \qquad (12.1.3)$$

or, by integrating over a short time interval, we can write conservation of energy as

$$dW = v\,dq - f\,dx \qquad (12.1.4)$$

This conservation equation can be valid only if energy is truly conserved inside the device. The presence of resistive losses will invalidate the result, so all lossy elements should be removed from consideration. Electrical losses can usually be included in the external circuitry which drives the device, and mechanical losses, such as friction, can usually be transferred to the model of the external mechanical system. If such a separation of losses is not possible, the device must be treated as a continuum, and none of the lumped methods of this section will be reliable. The same caveat applies to systems which contain energy sources, since these also will

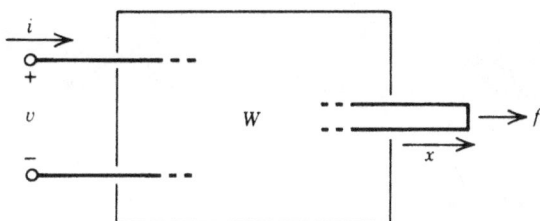

FIGURE 12.1.1. Energy flows to and from a device.

violate conservation of energy. Electrets, for example, contain a source of charge which, if not separated from the conservative system, can give erroneous results.

Once the conservative nature of the device is established, it is clear that the stored energy can depend only on the values of the terminal variables. A capacitor with given electrode spacing x and charge q will always have a definite energy, which depends on these variables,

$$W = W(q, x) \tag{12.1.5}$$

Changes in this energy are given by the chain rule of differentiation as

$$dW = \frac{\partial W}{\partial q}\bigg|_x dq + \frac{\partial W}{\partial x}\bigg|_q dx \tag{12.1.6}$$

The equation is very similar to conservation of energy, Eq. (12.1.4), but it has independent validity since it comes from a mathematical definition and not from a physical principle. Since both equations are true, and the independent variables x and q can be selected at will, the corresponding coefficients of the two equations must be identical. In particular, if we hold the charge constant ($dq = 0$) and let x vary, equating the two expressions for energy change gives

$$f = -\frac{\partial W}{\partial x}\bigg|_q \tag{12.1.7}$$

This is the expression we use to calculate the forces exerted by a lumped electrostatic device.

Before this expression can be used, the stored energy must be known. The easiest way to obtain this energy involves integration of the conservation equation in the form

$$W = \int (v\,dq - f\,dx) \tag{12.1.8}$$

Since there are two independent variables, the integral must be carried out as a line in two dimensions, as shown in Figure 12.1.2. In this figure the origin represents the reference from which all energies are determined. Since the energy is a conservative (or state) function, the integrals along any of the paths shown in the figure will give the same energy values. In practice, however, the integrals can not

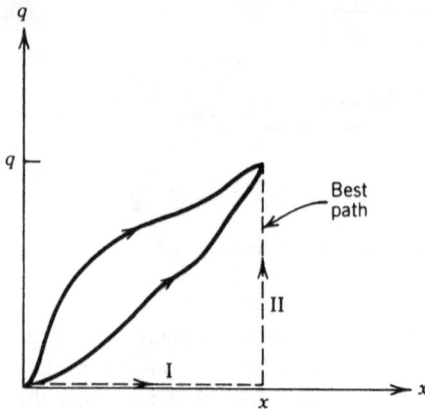

FIGURE 12.1.2. Paths of integration for energy.

be carried out along any arbitrary line since the integrand involves the force, which is not yet known.

Fortunately, there is a special path for electrostatic devices which allows us to find the energy without knowing the force. This is the rectangular path labeled "best" in Figure 12.1.2. Physically this path represents a sequence of first assembling the capacitor in its final configuration and then charging it. During the initial assembly (labeled I) the energy integral is

$$\Delta W_{\mathrm{I}} = \int_{q=0,x=0}^{q=0,x} (v\,dq - 0\,dx) = 0 \qquad (12.1.9)$$

Since the electrostatic force vanishes when the charge vanishes, the force is zero during the assembly, and no mechanical work is done. In addition, the charge is held constant (at zero), so no electrical power flows into the device. On the second leg of the path (labeled II) charge is added, and the electrostatic force builds up but motion is prevented, which precludes mechanical work. The only contribution to stored energy during this operation comes from the electrical input,

$$\Delta W_{\mathrm{II}} = \int_{\substack{q=0 \\ x=\mathrm{const}}}^{q} v\,dq \qquad (12.1.10)$$

Combining the contributions to the integral from both legs gives the total energy at the point x, q as

$$W = \Delta W_{\mathrm{I}} + \Delta W_{\mathrm{II}} = \int_{q=0}^{q} v(q, x = \mathrm{const})\,dq \qquad (12.1.11)$$

Note that this integral is the vestige of the complete line integral which grew out of conservation of energy. It must follow the constraints imposed by the special path to be valid. In particular, it must be carried out with x held constant. The integrand is the voltage on the capacitor, which is usually obtained from the terminal relation of the device. Most of the terminal relations, such as those discussed earlier in Chapter 11, come in the form

$$q = q(v, x) \qquad (12.1.12)$$

and must be solved to give the energy integrand as

$$v = v(q, x) \qquad (12.1.13)$$

before the integral can be carried out.

When using this method, it is rarely necessary to retrace all of the steps described unless the device has some unusual characteristics. Instead, the force expression is obtained directly from the electrical terminal relation by integrating it to find the energy and then differentiating to find the force.

Example: Electrostatic Loudspeaker

The electrostatic loudspeaker is a high quality, but expensive, alternative to the electromagnetic speakers used in most stereo systems. In its simplest form it consists of a moveable conducting membrane adjacent to an electrode, as shown in Figure 12.1.3. The voltage between the membrane and the electrode is controlled by the output of the audio amplifier and varies according to the audio signal. This voltage variation causes a variation in the force attracting the membrane, which in turn causes it to move, along with the air in contact with it. Typically, the voltage has both a bias and a signal component of the form

$$v = V_0 + V_1 \cos \omega t \qquad (12.1.14)$$

To design such a loudspeaker, it is necessary to know how much voltage must be applied to generate a force large enough to create an audible sound wave.

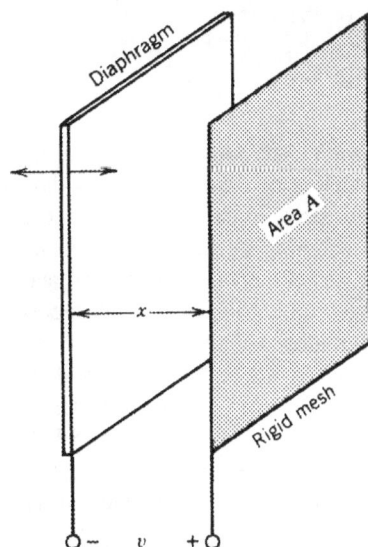

FIGURE 12.1.3. An electrostatic speaker.

Application of the Theory

The force can be calculated by using the energy method described previously. The first requirement is the terminal relation, which in this case corresponds to a simple parallel plate capacitor,

$$q = \frac{\epsilon A v}{x} \tag{12.1.15}$$

Solving for v gives

$$v = \frac{xq}{\epsilon A} \tag{12.1.16}$$

The energy integral becomes

$$W = \int \left(\frac{xq}{\epsilon A}\right) dq = \frac{xq^2}{2\epsilon A} \tag{12.1.17}$$

and differentiating gives the force as

$$f = -\left.\frac{\partial W}{\partial x}\right|_q = -\frac{q^2}{2\epsilon A} \tag{12.1.18}$$

The force is expressed as a function of the charge, but the amplifier normally applies a voltage to the speaker. We can easily express the force in terms of voltage by replacing q by the terminal relation to get the equivalent force expression

$$f = -\frac{\epsilon A v^2}{2x^2} \tag{12.1.19}$$

This is a general expression, valid for any combination of x and v. With the amplifier output given previously, the force takes on the form

$$f = \frac{\epsilon A}{2x^2}\left[\left(V_0^2 + \frac{V_1^2}{2}\right) + 2V_0 V_1 \cos \omega t + \frac{V_1^2}{2} \cos 2\omega t\right] \tag{12.1.20}$$

There are several noteworthy aspects of this result. The force depends on the separation between membrane and electrode x, which is constantly changing during the operation of the loudspeaker. When reproducing loud signals, this effect is greatest, and since it represents a distortion of the incoming signal, it represents an unwanted type of operation. In addition, the force has components at DC, ω, and 2ω, even though the input signal only has a single component at ω. This represents a harmonic distortion, which arises in the nonlinear voltage response of the device. Again, this is an effect which should be avoided in a high quality speaker.

The actual value of the force to be expected in an electrostatic speaker can be estimated from this expression. Taking the typical values of $V_0 = V_1 = 5$ kV, $A = 0.3$ m^2, $x = 5$ mm gives a force magnitude of approximately 2.7 N, or a pressure of $p = 8.9$ Pa. This pressure, along with the acoustic impedance of air

$(Z_a = 428$ kg/m-s), implies a power output on the order of

$$\frac{dW}{dt} = \frac{1}{2} p \cdot \left(\frac{p}{Z_a}\right) \approx 93 \text{ mW/m}^2$$

or 110 dB over the standard acoustic reference of 10^{-12} W/m^2.

Discussion

Two different expressions were obtained for the force on the membrane, corresponding to the use of voltage or of charge as the independent variable. Either of these expressions gives the correct value of force in terms of the position and the electrical terminal variables. In practice, however, the existence of two alternate expressions may lead to faulty conclusions concerning the behavior of a device. If the expression in terms of charge, Eq. (12.1.8), is plotted as a function of x, the curve marked a in Figure 12.1.4 is obtained, whereas if the expression in terms of voltage, Eq. (12.1.19), is plotted, the curve b results. One seems to imply a force independent of electrode spacing, whereas the other corresponds to a force which increases as the plates approach. Clearly, both can not be right.

In fact, neither of the curves is necessarily correct. Both of them entail specific assumptions concerning the external electrical circuit attached to the device. One assumes that the voltage is held constant; the other assumes constant charge on the plates (open circuit). In the example just described, for example, neither the voltage nor the charge remains constant during the operation, so neither curve is appropriate.

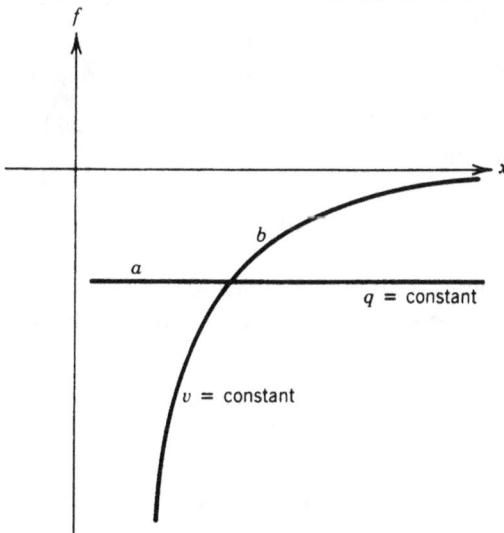

FIGURE 12.1.4.   Position dependence of alternate force expressions.

12.2   TORQUE CALCULATION
(Electrostatic Display Signs)

Summary

Electrostatic torques can be calculated with the energy method just as easily as linear forces. The energy integral is carried out by using only the electrical terminal relation, and the torque is obtained by a simple derivative. The method is illustrated by a digitally driven electrostatic sign, which also shows how clever design can compensate for the relatively weak forces commonly encountered in electrostatic devices.

Theory

The energy method described in the previous section is not limited to linear forces. It can just as easily be used to find the torques generated by an electrostatic device. The only change comes in the nature of the mechanical work term in the conservation of energy. When the mechanical work is done by rotation, conservation of energy can be written as

$$dW = v\,dq - \tau\,d\theta \tag{12.2.1}$$

Energy is still a state function, so the differential can also be written as

$$dW = \left.\frac{\partial W}{\partial q}\right|_{\theta} dq + \left.\frac{\partial W}{\partial \theta}\right|_{q} d\theta \tag{12.2.2}$$

Comparing terms in the same manner as before, the torque is given by

$$\tau = -\left.\frac{\partial W}{\partial \theta}\right|_{q} \tag{12.2.3}$$

As before, the energy can be calculated by a two-dimensional line integral derived from the conservation equation,

$$W = \int (v\,dq - \tau\,d\theta) \tag{12.2.4}$$

which is best taken over a path similar to that shown in Figure 12.1.2 so that the torque does not have to be known in advance. Using this path, the energy is given by

$$W = \int_{q=0}^{q} v(q, \theta = \text{const})\,dq \tag{12.2.5}$$

Example: An Electrostatic Display Sign

Digital displays driven by computers are common now, but most of these are relatively small devices intended to be used at a desktop level. Large advertising signs can also be computer driven, but these devices are often expensive because of the large areas involved. An electrostatic display sign (Kalt, 1975) has been developed to offer a less expensive alternative to incandescent lights and other devices used on a large scale.

The basic mechanism of the sign is shown in Figure 12.2.1. The vertical electrode is painted black, while the moveable conducting vane is painted white. With the voltage turned off, a spring pulls the white vane down so that the observer sees only the black vertical electrode. Application of voltage to the electrodes generates a torque which pulls the white vane up, covering the black background. This represents one pixel of a larger display, with each pixel controlled by a separate voltage source. The basic design goal is to determine how much voltage the driver circuit must supply to rotate the vane.

Application of the Theory

The torque can be calculated from the energy method, beginning with the terminal relation of the device. The electric field between the electrodes is perpendicular to both electrodes, and must therefore be directed in the θ direction. Solution of the field equations gives the electric field as

$$E_\theta = \frac{v}{r\theta} \tag{12.2.6}$$

The charge on the positive electrode is given in the usual fashion by integration over the electrode surface as

$$q = \int \mathbf{D} \cdot d\mathbf{A} = \epsilon d \, \ln\left(\frac{b}{a}\right) \frac{v}{\theta} \tag{12.2.7}$$

Solving for the voltage and integrating to find the energy gives

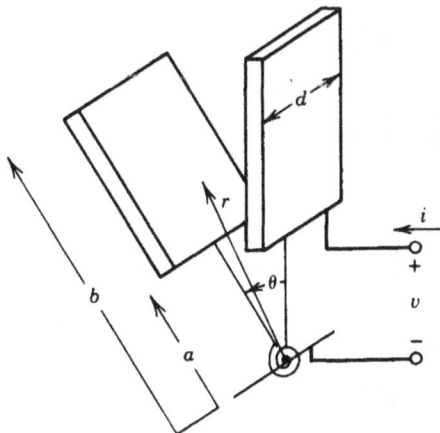

FIGURE 12.2.1. Electrostatic sign mechanism.

$$W = \int v\,dq = \frac{\theta q^2}{2\epsilon d\,\ln(b/a)} \tag{12.2.8}$$

Finally, differentiation gives the torque as

$$\tau = -\left.\frac{\partial W}{\partial \theta}\right|_q = -\frac{q^2}{2\epsilon d\,\ln(b/a)} = -\frac{\epsilon d\,\ln(b/a)v^2}{2\theta^2} \tag{12.2.9}$$

An estimate of the torque produced for some typical dimensions is necessary for designing such a sign. Assuming $d = 2$ cm, $b/a = 10$, $\theta = 0.5$ rad, $v = 10$ kV, gives a torque of approximately 0.082 mN-m in air. This is a very small torque, and it requires switching of high voltages at the information transmission rate, so this particular configuration is not likely to be practical.

Discussion

The relatively small forces in this simple example are typical of electrostatic devices and might at first discourage potential users of these effects. Often, however, clever design can magnify the effect and lead to a practical device. In the sign application, for example, digitally controlled signs with driver voltages on the order of 100 V have been successfully developed (Kalt, 1975) by changing the configuration. Instead of a rigid electrode which must be attracted from some distance away, a flexible electrode curled away from the fixed electrode is employed, as shown in Figure 12.2.2. The torque is much stronger near the point where the electrodes come together, so a much smaller voltage is needed to bring that portion of the flexible electrode up to cover the fixed one. Once the lower portion is in contact, the torques continue to act on the upper portion, so that eventually the entire flexible electrode has unwrapped to cover the fixed one and change the color of that pixel.

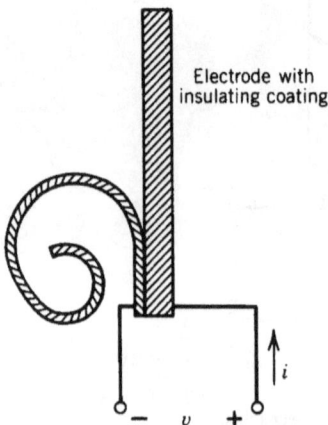

Electrode with insulating coating

FIGURE 12.2.2. A more practical electrostatic sign.

12.3 FORCES FROM COENERGY
(Ferroelectric Forces)

Summary

When an electrostatic device has a complicated terminal relation, the energy method often fails at the step where inversion of the terminal relation is required. This inversion can be avoided by an alternate method based on coenergy, which is illustrated by the calculation of force for a ferroelectric transducer.

Theory

In each of the examples of energy given in the last two sections it was necessary to invert the terminal relation to express voltage as a function of charge before doing the energy integral. This is a slight annoyance for the linear devices considered so far, but it can become a serious problem when the device has a nonlinear terminal relation, or a more complex linear one involving many terminals. Fortunately, there is an alternate approach to the energy integral which sidesteps the inversion of the terminal relation.

The new approach involves a different energy function, called the coenergy. The coenergy method (Woodson and Melcher, 1968) begins from conservation of energy, just as the energy method does. For a simple capacitor with linear motion this can be written as

$$dW = v\,dq - f\,dx \qquad (12.3.1)$$

as before. As soon as the conservation equation has been written in this form, the independent variables are automatically fixed. In this case x and q must be used since they appear as differentials, and the subsequent integration must be performed in the $q-x$ plane. This integration implies an integrand in the form $v(q, x)$, which requires inversion of the terminal relation. It would be more convenient if it could be performed in the $v-x$ plane with $q(v, x)$ as the integrand.

Fortunately, it is always possible to interchange the two terminal variables by using the chain rule for differentials. In this case we can write the offending term as

$$v\,dq = d(vq) - q\,dv \qquad (12.3.2)$$

Substituting back into the conservation equation and collecting terms gives

$$d(qv - W) = q\,dv + f\,dx \qquad (12.3.3)$$

This technique, called a Legendre transformation, yields an equation very similar to the conservation of energy and in fact represents a conservation law for the related quantity

$$W' = qv - W \qquad (12.3.4)$$

which is called the coenergy. Just like the energy, the coenergy is a conservative (or state) function, so its differential can be written in terms of the terminal variables

$$dW'(v, x) = \frac{\partial W'}{\partial v}\bigg|_x dv + \frac{\partial W'}{\partial x}\bigg|_v dx \qquad (12.3.5)$$

Comparing terms of this equation with the coenergy conservation, Eq. (12.3.3) leads to an alternate formulation of the force expression

$$f = +\frac{\partial W'}{\partial x}\bigg|_v \qquad (12.3.6)$$

Although this force expression is superficially similar to that involving the energy, there are two critical differences. The sign of the expression is positive, whereas it was negative for the energy. This means that the force direction will be reversed if the wrong energy function is used. A second point of dissimilarity appears when taking the partial derivative. Here, for the coenergy method, the voltage must be held constant, whereas in the energy method the charge was held constant. Again, using the wrong energy function can give an incorrect answer.

The coenergy function is found by using the same procedure as for the energy function, namely, a line integral over a rectangular path. In this case the path is carried out in the v–x plane shown in Figure 12.3.1. The path is taken in the x direction first, with the voltage turned off so that no force is generated and no co-work added to the device. The path finishes along the vertical leg, while mechanical motion is prevented so that no mechanical co-work can be done. This leaves only the electrical co-work integral,

$$W' = \int_{v=0}^{v} q(v, x = \text{const})\, dv \qquad (12.3.7)$$

The integrand here is given directly by the terminal relation, so no inversion is needed.

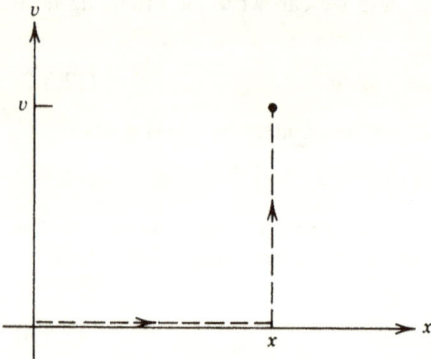

FIGURE 12.3.1. Path for coenergy integration.

Example: Ferroelectric Forces

Forces exerted by nonlinear dielectrics such as ferroelectrics are a natural example for the use of coenergy, rather than energy. The nonlinearity in the material is a reflection of the increased interaction available with such materials, so it is likely to be a factor in any device which relies on this interaction to increase its output.

The relation between the D field and the E field in these materials is usually quite complicated, but the technique can be illustrated with a piecewise linear approximation to the constitutive relation of the form

$$D = \epsilon E, \qquad E < \frac{D_0}{\epsilon} \qquad (12.3.8a)$$

$$D = D_0, \qquad E > \frac{D_0}{\epsilon} \qquad (12.3.8b)$$

In a parallel plane geometry, with area A and separation x, the electric field is given by

$$E = \frac{v}{x} \qquad (12.3.9)$$

and the terminal relation takes the form

$$q = \frac{\epsilon A v}{x}, \qquad v < \frac{x D_0}{\epsilon} \qquad (12.3.10a)$$

$$q = A D_0, \qquad v > \frac{x D_0}{\epsilon} \qquad (12.3.10b)$$

which is shown in Figure 12.3.2. Note that the breakpoint voltage depends on the electrode spacing, which may change during the operation and in any case must remain variable to carry out the derivative which gives the force.

Application of the Theory

Now that the terminal relation is known, the coenergy can be evaluated by integration. There are two cases to be considered, depending on whether the voltage exceeds the critical value which saturates the dielectric. Below saturation (the double-hatched region in Fig. 12.3.2),

$$W' = \int q \, dv = \frac{\epsilon A v^2}{2x} \qquad (12.3.11)$$

and

$$f = \frac{\partial W'}{\partial x}\bigg|_v = -\frac{\epsilon A v^2}{2x^2}, \qquad v < \frac{x D_0}{\epsilon} \qquad (12.3.12)$$

Above saturation, the coenergy integral splits into two parts as the voltage is

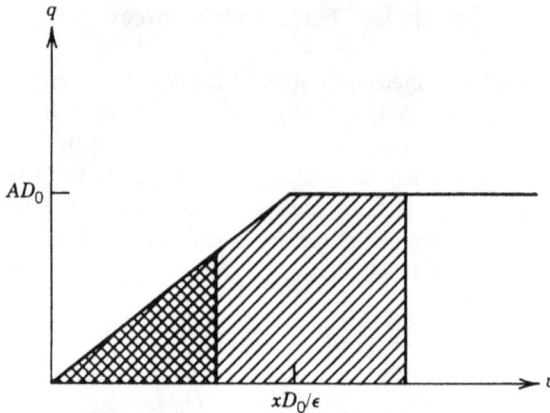

FIGURE 12.3.2.   Ferroelectric terminal relation.

raised from zero past the saturation point to its final value. In this case

$$W' = \int_{v=0}^{xD_0/\epsilon} \frac{\epsilon A v}{x}\, dv + \int_{v=xD_0/\epsilon}^{v} AD_0\, dv$$

$$= AD_0 v - \frac{AD_0^2 x}{2\epsilon} \tag{12.3.13}$$

and the force is

$$f = \left.\frac{\partial W'}{\partial x}\right|_v = -\frac{AD_0^2}{2\epsilon}, \qquad v > \frac{xD_0}{\epsilon} \tag{12.3.14}$$

The breakpoint caused by saturation affects the force as either voltage or spacing changes. If the spacing is fixed and the voltage increased, the magnitude of the force, which is shown in Figure 12.3.3a, increases quadratically until saturation, but it gets no larger above that point. Thus there is no advantage to operating at voltages above saturation.

The effect of spacing also shows significant differences on either side of the saturation point, as shown in Figure 12.3.3b. If the electrode spacing is small, the dielectric is saturated, and the force remains constant. At larger spacing, the dielectric is no longer saturated, and the force falls off at higher separations. Thus the ferroelectric transducer has two distinctly different operating modes, depending on the state of the dielectric saturation.

Discussion

The coenergy method is used more often than the energy method in practice because it eliminates the inversion of the terminal relation, making it simpler in every case. Because of its frequent use, the distinction between it and the energy method is often lost in casual conversation, and mistakes sometimes follow as a result of sign changes or partial derivatives taken while holding the wrong variable

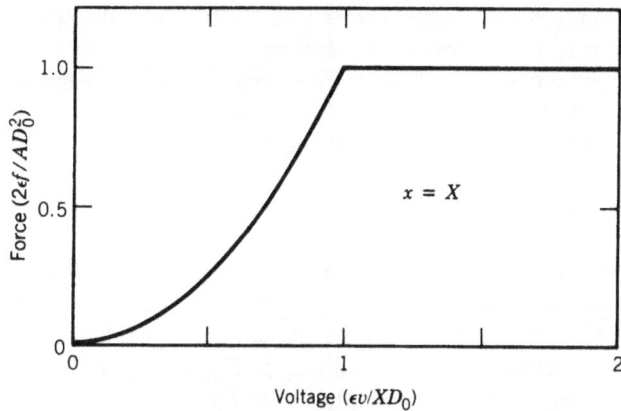

(a)

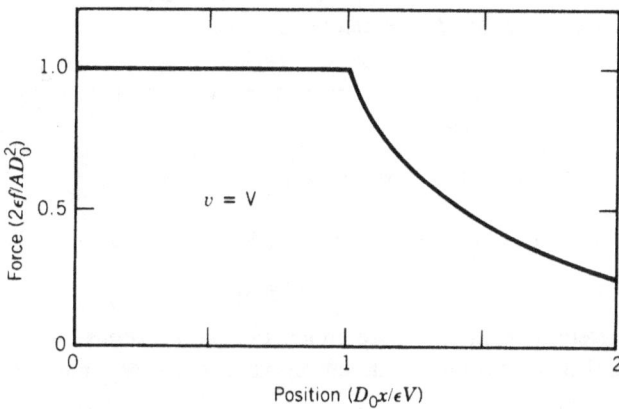

(b)

FIGURE 12.3.3. Saturation effects on the force produced by a ferroelectric transducer.

constant. It turns out that the sign change is often tacitly ignored by arbitrarily changing it to give the physically expected result, whereas the magnitude always comes out the same if the terminal relation is linear and simple. Thus this confusion causes little difficulty with a simple system, but it will introduce serious errors in a nonlinear system.

12.4 FORCES IN MULTIPLE-TERMINAL DEVICES
(Electrostatic Motors)

Summary

Multiple-terminal electrostatic devices imply a reference (or ground) potential which can be selected at will by the designer. Once selected, however, it restricts the operating range, so its selection is best left to the end of the analysis, where it should be explicit. The forces (or torques) in such devices can be calculated by means of an energy function before the reference voltage has been selected by using extensions of the procedures introduced previously. A rotating electrostatic motor makes a good example of this technique.

Theory

The energy method can also be used when the device has multiple terminals, but some of the characteristics of the voltage reference level must be kept clear to avoid confusion. The simplest multiple-terminal device which illustrates the technique is the three-terminal device illustrated in Figure 12.4.1. In addition to the three terminal voltages shown explicitly in the figure, there is always a voltage reference which can be chosen arbitrarily, since only potential differences have a direct physical meaning. For example, taking the reference voltage as v_{ref}, the power flow equation for this device can be written as

$$\frac{dW}{dt} = (v_1 - v_{ref})i_1 + (v_2 - v_{ref})i_2 + (v_3 - v_{ref})i_3 - f\frac{dx}{dt}$$

$$= (v_1i_1 + v_2i_2 + v_3i_3 - v_{ref}(i_1 + i_2 + i_3) - f\frac{dx}{dt} \tag{12.4.1}$$

Because there is no internal source of charge (this is a conservative device),

$$i_1 + i_2 + i_3 = 0$$

and the power term containing the reference voltage v_{ref} drops out of the power flow equation. This means that the reference voltage can be arbitrarily set to any

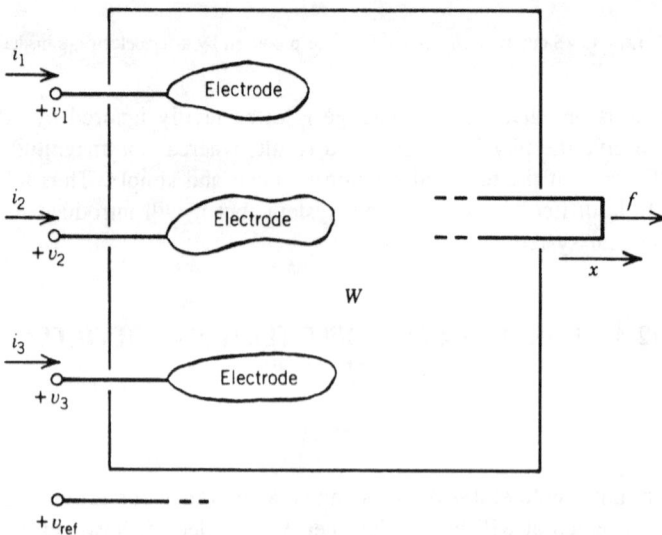

FIGURE 12.4.1. An electrostatic device with three terminals.

value which is convenient for the task at hand. One common choice is to select this reference voltage so that one of the terminal voltages becomes zero. For example, if the reference voltage is selected so that $v_3 = 0$, the power flow equation simplifies to

$$\frac{dW}{dt} = v_1 i_1 + v_2 i_2 - f \frac{dx}{dt} \tag{12.4.3}$$

Another common choice involves balanced excitation, where the voltages on the three electrodes are related by a simple symmetry, such as three-phase sinusoidal voltages with phase angles 120° apart. In this case, of course, all three voltages will appear in the power expression.

Once an appropriate power expression is selected, the energy method can be applied to find the forces. Using the simplest case as an example, with the reference chosen so that $v_3 = 0$, the energy differential becomes

$$dW = v_1 \, dq_1 + v_2 \, dq_2 - f \, dx \tag{12.4.4}$$

The terminal relations obtained with multiple-terminal devices are always fairly complicated, so the coenergy method is usually selected to avoid the inversion of terminal relations required by the energy method, as discussed in the previous section. Applying the Legendre transformation gives the coenergy differential as

$$dW' = q_1 \, dv_1 + q_2 \, dv_2 + f \, dx \tag{12.4.5}$$

Since the coenergy is a state function, its differential can also be written in terms of partial derivatives as

$$dW' = \frac{\partial W}{\partial v_1} \, dv_1 + \frac{\partial W'}{\partial v_2} \, dv_2 + \frac{\partial W'}{\partial x} \, dx \tag{12.4.6}$$

and the force is again given by

$$f = \frac{\partial W'}{\partial x} \bigg|_{\substack{v_1 \\ v_2}} \tag{12.4.7}$$

The coenergy is obtained by integrating the conservation equation (12.4.5) along a rectangular path in the v_1–v_2–x space shown in Figure 12.4.2.

As in the previous energy integrals, there will be no contribution from mechanical work along path I because the force will be zero until a voltage is turned on. Along path II, as v_1 is increased, energy will be introduced into terminal 1 only, giving a coenergy increment of

$$W' = \int_{v_1=0}^{v_2} q_1(x = \text{const}, v_1, v_2 = 0) \, dv_1 \tag{12.4.8}$$

while along the last leg (III) coenergy increases by

$$W' = \int_{v_2=0}^{v_2} q_2(x = \text{const}, v_1 = \text{const}, v_2) \, dv_2 \tag{12.4.9}$$

Note that different constraints are imposed on the two integrals. Along path II, v_2

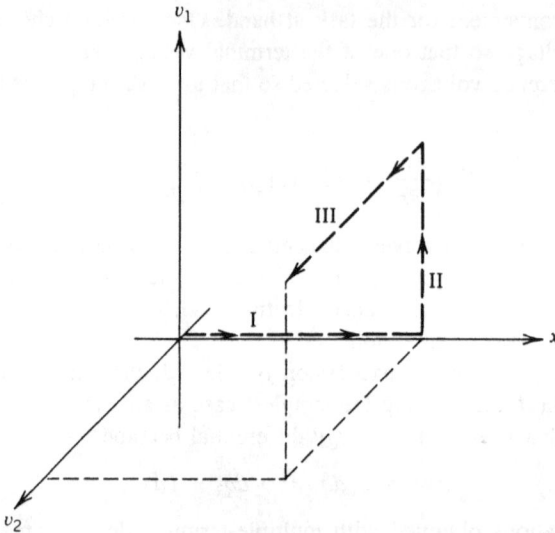

FIGURE 12.4.2. Integration path for coenergy with multiple terminals.

vanishes, while along path III, v_1 is held at a fixed, nonzero value. Thus the two integrals are not symmetrical. The total coenergy is obtained by adding the contributions from each path, and the force then follows by differentiation, according to Eq. (12.4.7).

Example:　An Electrostatic Motor

Electrostatic motors can take many forms, but many of them can be represented by a relatively simple model, such as that shown in Figure 12.4.3. The motor con-

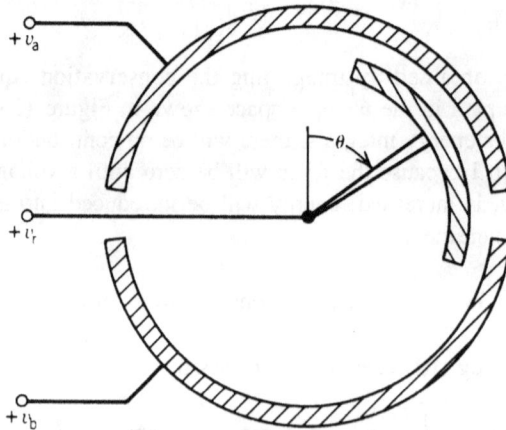

FIGURE 12.4.3. A simple electrostatic motor.

sist of two stationary electrodes, or stators, called the a and the b stators, and one rotating electrode, or rotor, at a position θ. The terminal relations for this motor can be calculated using the standard techniques described in Section 11.4, but in the interests of simplicity, we will take approximate expressions of the form

$$q_a = (C_{s0} - C_{s1} \cos \theta)v_a - (C_{m0} - C_{m1} \cos \theta)v_r \tag{12.4.10}$$

$$q_b = (C_{s0} + C_{s1} \cos \theta)v_b - (C_{m0} + C_{m1} \cos \theta)v_r \tag{12.4.11}$$

$$q_r = C_r v_r - (C_{m0} - C_{m1} \cos \theta)v_a - (C_{m0} + C_{m1} \cos \theta)v_b \tag{12.4.12}$$

This form involves just the leading terms in the Fourier series which describe the self- and mutual capacitances at the three electrodes. Simplifications like this are often used in the study of rotating machines. Since we would like a torque expression which is valid for any type of electrical excitation, no assumptions have yet been made about the voltage reference.

Since there are three electrical terminals, the coenergy integral will involve three contributions, which will have the forms

$$\Delta W_I = \int_{v_b = v_r = 0} q_a \, dv_a = \frac{1}{2}(C_{s0} - C_{s1} \cos \theta)v_a^2 \tag{12.4.13}$$

$$\Delta W_{II} = \int_{v_r = 0} q_b \, dv_b = \frac{1}{2}(C_{s0} + C_{s1} \cos \theta)v_b^2 \tag{12.4.14}$$

$$\Delta W_{III} = \int q_r \, dv_r = \frac{1}{2}C_r v_r^2 - (C_{m0} - C_{m1} \cos \theta)v_a v_r$$
$$- (C_{m0} + C_{m1} \cos \theta)v_b v_r \tag{12.4.15}$$

Adding these terms gives the total coenergy, which can then be differentiated to give the torque as

$$\tau = \tfrac{1}{2}C_{s1} \sin \theta \, (v_a^2 - v_b^2) - C_{m1} \sin \theta \, v_r(v_a - v_b) \tag{12.4.16}$$

Various types of electric machines can be described by setting the terminal constraints corresponding to different types of operation. For example, if the stators are driven by opposite terminals of a voltage source $v_a = -v_b = v_s/2$, a synchronous machine results, with a torque expression

$$\tau = -C_{m1}v_r v_s \sin \theta \tag{12.4.17}$$

Other machines may be implemented with different terminal constraints. Some of these are described in the following problems.

Discussion

It should be clear from the foregoing example that it is not necessary to rush into a choice of the reference voltage when describing a multiterminal device. In fact, once the choice is made, the results are always restricted to the simplest case, which may not be particularly useful in application. For example, a natural choice for the reference voltage might correspond to a grounded rotor $v_r = 0$, since many rotating electrostatic motors operate with this constraint. In this particular machine, however, grounding the rotor completely eliminates the torque [given in Eq. (12.4.18)]. Thus an important class of devices would be completely overlooked by a premature choice of a reference voltage.

The additional terminals present no real problem in force calculations since the coenergy method can always be extended by additional line segment integrals. In practice, electrostatic devices such as motors may have dozens of individual electrodes, but the torque can always be obtained by following the procedures given here, with straightforward extensions, so long as the machine can be modeled as a conservative system.

BIBLIOGRAPHY

Crowley, J. M., Electrical breakdown of bimolecular lipid membranes as an electromechanical instability, *Biophys. J.*, **13:** 711–724 (1973).

Greason, W. D., A modular changeable message electrostatiac display system, *IEEE-IAS Annu. Meet. Conf. Rec.*, 1984, pp. 1038–1044.

Jefeminko, O. D., Electrostatic motors, in *Electrostatics and its Applications*, A. D. Moore, Ed., Wiley, New York, 1973, Chap. 7.

Jones, T. B., Lumped parameter electromechanics of electret transducers, *IEEE Trans. Acoustics, Speech and Signal Processing*, **ASSP-22:** 141–145 (1974); **ASSP-23:** 497–501 (1975).

Kalt, C., Electrostatic display device with variable reflectivity, U.S. Patent 3,897,997, August 5, 1975.

Woodson, H. H., and J. R. Melcher, *Electromechanical Dynamics*, Vol. 1, Wiley, New York, 1968, Chap. 3.

PROBLEMS

PROBLEM 1   (INSTRUMENTATION)

An electrostatic voltmeter consists of a moveable electrode of width d ($= 10$ cm) which translates parallel to two fixed electrodes spaced a distance a ($= 2$ mm) away. The moveable center electrode is restrained by a spring which applies a force $f = K(x - x_0)$ which maintains the overlap between fixed and moveable electrodes at x_0. Find the deflection of the center plate in terms of the applied voltage v ($= 10$ kV).

PROBLEM 2   (BIOMEDICAL)

Cells can be joined by electrofusion, a process which plays a growing role in genetic engineering. When a cell membrane is subjected to an electrostatic field, the compression forces on the membrane can cause pore formation. The behavior of the membrane during electrofusion can be modeled by a parallel plate capacitor with electrode separation x containing a dielectric material ($\kappa = 3$) which acts as an elastic spring holding the outer surfaces of the membrane apart at a distance l ($= 1$ nm) with a force $f = K(x - l)$.

    a. Find the equilibrium thickness of the membrane when a voltage V ($= 1$ V) is applied.

    b. What is the largest voltage that can be applied before the electric forces overcome the elastic forces and the membrane ruptures?

PROBLEM 3 (OFFICE EQUIPMENT)

In the ink jet printer described in Section 11.4, the drop which is just forming is subject to a force caused by the charge on the previously formed drop. Find an expression for this force.

PROBLEM 4 (POWER)

Alone on the desert, an engineer discovers that his car battery is too weak to turn the engine, although the open circuit voltage is still at its rated value V_0. He decides to recharge the battery using the tuning capacitor of his car radio as an energy converter and operating the battery–capacitor system over the closed cycle shown in Figure 12.P.4.

The capacitor has the terminal relation $q = C_0 v \sin \theta$.

    a. Find an expression for the torque on the capacitor shaft.

    b. In which direction must the path $abcd$ be traversed to charge the battery?

    c. How much energy is put into the battery in one cycle, assuming the conditions of part b?

    d. Using the values $V_0 = 12$ V and $C_0 = 360$ pF, find the number of cycles needed to recharge the battery by 10^5 J.

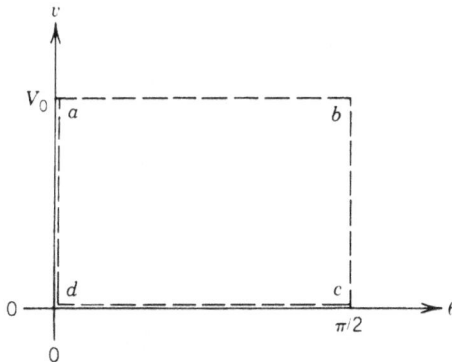

FIGURE 12.P.4. Energy conversion cycle.

PROBLEM 5

An electret motor has the terminal relations

$$q_s = C_s v_s + C_m v_r \cos \theta, \qquad q_r = C_r v_r + C_m v_s \cos \theta$$

To put the machine into operation, the electret rotor is first polarized at $\theta = 0$ by applying a voltage $v_r = V_0$ with the stator short circuited $v_s = 0$. From now on, the rotor will remain open circuited ($i_r = 0$) to conserve the charge on the electret. During operation the stator is driven with a voltage source $v_s = V_0 \cos \omega t$, and the rotor turns at a steady speed so that $\theta = \omega t + \theta_0$. Find the torque produced by the motor.

AREAS OF APPLICATION

When does electrostatics concern an engineer? One obvious situation involves the appearance of high voltages. Electric power transmission, X-ray machines, and lightning protection are obviously associated with strong electric fields and will require electrostatics to understand and design suitable equipment. Although these examples all involve high voltages, it is really the high electric fields, not the voltages in themselves that are responsible for the effects. Even higher electric field strengths can be reached in smaller volumes, where the total voltage drop may be deceptively small. An ordinary cell in the human body supports a potential difference of approximately 0.1 V across its membrane, but the electric field which stresses the membrane is three times that needed to form sparks in air. In fact, relatively high electric fields are much less likely to lead to breakdown in small devices, a fact which tends to favor electrostatics when small objects must be controlled or manipulated.

Electrostatic forces are usually considered too weak to move macroscopic objects and give way to magnetic forces when large, powerful systems are needed. For most of the last century, as motors and generators were called on to replace the muscles of humans and animals, magnetic forces and interactions were a principal concern of engineers. In the last decade, however, it has become clear that engineers must now be concerned with replacing brains, not muscles, a task which requires speed rather than strength. High speeds imply small sizes to minimize inertia, and small size implies that electrostatic effects will become important. Engineers are concerned now with moving electrons, not mills, and this is accomplished by the electrostatic fields within solid-state electronic devices such as transistors, MOSFETs (metal oxide semiconductor field effect transistors), and computer chips. Even at the larger scale involved in input and output devices, electrostatics plays a dominate role in cathode ray tube (CRT) and liquid crystal displays and in output printers. In these devices high resolution calls for small sizes, so the higher quality output devices tend to use electrostatic effects.

Electrostatic effects do not just occur in engineered systems; they arise in nature also. The atmosphere contains numerous charged particles which are involved in all kinds of precipitation, including the most impressive example, the thunderstorm. The basic forces of physics and chemistry involve electrostatics, making it impossible to discuss atoms or molecules without a good understanding of electrostatics. Even in walking across a room, a person can generate electrostatic voltages of 20 kV, which are often unpleasant and may be diastrous for electronic circuits.

High voltage work and basic science have always required electrostatics, but its widespread use in engineering had to await the development of cheap, reliable, and effective insulators. Without them, charge must be continually resupplied to maintain the high electric fields, and the requirements of a high voltage power supply made many devices impractical. High quality insulators are now available, and they have made possible many new electrostatic devices, such as the field effect transistor (FET) and the walk-around headphone. Good insulators also appear in many other places, such as polyester clothing, where they enhance the effects of static electricity, leading to gate breakdown in the FETs during manufacture and assembly. Thus electrostatics is a two-edged sword, and its behavior is of interest to many more than those who seek to use it.

Electrostatics has grown up in several areas of application, each with its own peculiar combination of constraints and operating conditions. In some places, such as electric power lines, only the basic potential fields are needed. In others, such as solid-state electronics, effects like diffusion are important, whereas in CRTs inertia plays a prominent role. The preceding chapters, which are arranged by increasing complexity of the phenomena encounted in the applications, may contain much information that is not relevant to a worker in a particular field. To help such a person find what parts of electrostatics are most relevant to the job at hand, a brief survey of the principal areas of application is included in this epilogue. The areas covered here are based on the industry classifications used in *Business Week's* annual report on research and development spending, with the addition of some areas normally supported by government and combinations or deletions of areas in which the electrostatics activity is relatively small. Each writeup includes a short description of some of the more important applications of electrostatics and ends with a bibliography of selected books and review articles. The areas are arranged alphabetically.

Inculet, I. I., Electrostatics in industry, *J. Electrostat.*, **4**: 175–192.

Masuda, S., Industrial applications of electrostatics, *J. Electrostat.*, **10**: 1–15 (1981).

Aerospace

Air- and spacecraft are isolated conductors in insulating media, so they will hold a net charge for very long times. This charging can become a serious problem when the craft must make contact with another object because the resulting sparking can destroy the electronics or ignite fuel vapors. Corona discharges can also occur from the craft, leading to interference with the radio and other electronics systems. Interference is also generated by electrostatic phenomena in the atmosphere, such as lightning. With the advent of composite nonmetallic structures, the effects of lightning have become more important since the conducting skin is no longer available to channel the stroke away from the interior. In addition to these effects peculiar to aerospace industry, many of the instruments, from fuel gauges to gyroscopes, depend on electrostatic effects for their operation. Finally, propulsion of spacecraft routinely uses electrostatic ion engines.

Inculet, I. I., Bi-ionized static discharger for aircraft, IEEE-IAS Annual Meeting Conference Record, 1979, pp. 109–111.

Jahn, R. G., *Physics of Electric Propulsion*, McGraw-Hill, New York, 1968.

Newman, M. M., and J. D. Robb, Protection of aircraft, in *Lightning*, Vol. 2, R. H. Golde, Ed., Academic, New York, 1977, Chap. 21.

Rittner, E. S., Recent advances in silicon solar cells for space use, *Adv. Electron. Electron Phys.*, **42:** 41–54 (1976).

Sodha, M. S., and S. Guha, Physics of colloidal plasmas, *Adv. Plasma Phys.*, **4:** 219–309 (1971).

Stuhlinger, E., and G. Mesmer, Eds., *Space Science and Engineering*, McGraw-Hill, New York, 1965.

Agriculture and Food Processing

Electrostatics appears at most places along the food chain, usually in a supporting role. It is used to sort seeds, to direct sprays to the plant, to measure the moisture content of the crop, to spin cotton, and to speed the baking of bread and the smoking of meat. It also appears as a hazard and has been blamed for some serious explosions at grain elevators. It plays a major role in the Russian fishing industry, where electric fields are often used to guide fish into nets.

Bazhenov, V. A., *Piezoelectric Properties of Wood*, Consultants Bureau, New York, 1961.

Cobine, J. D., Other electrostatic effects and applications, in A. D. Moore, Ed., *Electrostatics and Its Applications*, Wiley, New York, 1973, Chap. 9.

Eden, H. F., Electrostatic nuisances and hazards, in A. D. Moore, Ed., *Electrostatics and Its Applications*, Wiley, New York, 1973, Chap. 18.

Inculet, I. I., G. S. P. Castle, et al., Deposition studies with a novel form of electrostatic crop sprayer, *J. Electrostat.*, **10:** 65–72 (1981).

Inculet, I. I., G. S. P. Castle, and R. S. Vermey, Electrostatic spraying of row field crops, IEEE-IAS Annual Meeting Conference Record, 1984, pp. 1058–1060.

Inculet, I. I., et al., Spraying of electrically charged insecticide aerosols in enclosed spaces, *IEEE Trans. Ind. Appl.*, **IA-20:** 677–681 (1984).

Kulacki, F. A., and J. A. Daumenmier, A preliminary study of electrohydrodynamic augmented baking, *J. Electrostat.*, **5:** 325–336 (1978).

Law, S. E., Electrostatic pesticide spraying: concepts and practice, *IEEE Trans.*, **IA-19:** 160–168 (1983).

Lawver, J. E., and W. P. Dyrenforth, Electrostatic separation, in A. D. Moore, Ed., *Electrostatics and Its Applications*, Wiley, New York, 1973, Chap. 10.

Singh, S., and P. Cartwright, A study of electrostatic activity in grain silos, *IEEE Trans. Ind. Appl.*, **IA-20:** 863–868 (1984).

Sternin, V. G., *Electrical Fishing*, Israel Program for Scientific Translations, Jerusalem, 1976.

Tombs, M., Seed sorting, *La Physique des Forces Electrostatique at Leurs Applications*, Centre National de la Recherche Scientifique, Paris, 1961.

Atmosphere

Lightning is the best known electrostatic aspect of the atmosphere, but electric fields and charges are always present in the atmosphere and are involved in many precipitation processes in addition to thunderstorms. Since the ionosphere contains even more of these charges, understanding of its behavior involves electrostatics effects in greater scope.

Abbas, M. A., and J. Latham, The instability of evaporating charged drops, *J. Fluid Mech.*, **30**: 663–670 (1967).

Chalmers, J. A., *Atmospheric Electricity*, 2nd ed., Pergamon, Oxford, 1967.

Golde, R. H., Ed., *Lightning*, Vols. 1 and 2, Academic, New York, 1977.

Israel, H., *Atmospheric Electricity*, Vol. 1, Israel Program for Scientific Translations, NSF, Washington, DC, TT67-51394/1, 1971.

Mason, B. J., *The Physics of Clouds*, Clarendon, Oxford, 1971.

Rishbeth, H., and O. K. Garriott, *Introduction to Ionospheric Physics*, Academic, New York, 1969.

Schneider, J. M., N. R. Lindblad, and C. D. Hendricks, An apparatus to study the collision and coalescence of liquid aerosols, *J. Colloidal Sci.*, **20**: 610–616 (1965).

Tilson, S., Electricity and weather modification, *IEEE Spectrum*, April 1969, pp. 26–61.

Uman, M. A., *Lightning*, McGraw-Hill, New York, 1969.

Vonnegut, B., Atmospheric electrostatics, in *Electrostatics and Its Applications*, A. D. Moore, Ed., Wiley, New York, 1973, Chap. 17.

Whitten, R. C., and I. G. Poppoff, *Fundamentals of Aeronomy*, Wiley, New York, 1971.

Biomedical

Electrocution is the electrostatic effect of most concern to those who work with high voltages, but there are numerous other electrostatic effects and applications in biology and medicine. Electrophysiology is a fertile field for research since cells invariably possess large electric fields, and some, such as nerve cells, depend on these fields for their basic operation. Externally applied fields are known to affect the behavior of many cells and are used to fuse cells in genetic engineering.

In medical work, diagnosis is often carried out with the aid of electrostatics, as incorporated in electrocardiograms, electroencephalograms, and other recordings of organs with electrical activity including eyes, ears, and stomachs. *In vitro* diagnosis is aided by cell counters and sorters, electrophoresis, immunoelectrophoresis, pH measurements, and other instruments which are based on electrostatic principles. Electric therapy has had a long (and somewhat spotty) record in medicine. It is currently used in areas such as electrosurgery to prevent bleeding, in artificial nerve stimulation for some kinds of paralysis, and in aerosol therapy. It has also appeared in other forms, such as hair electrolysis, electroacupuncture, and shock therapy.

Adelman, W. J., *Biophysics and Physiology of Excitable Membranes*, Van Nostrand Reinhold, New York, 1971.

Cameron, J. R., and J. G. Skofronick, *Medical Physics*, Wiley, New York, 1978.

Cobbold, R. S., *Transducers for Biomedical Measurements*, Wiley, New York, 1974.

Dalziel, C. F., Electric shock, *Adv. Biomed. Eng.*, **3**: 223–245 (1973).

Geddes, L. A., *Electrodes and the Measurement of Bioelectric Events*, Wiley, New York, 1972.

Hughes, J. F., and A. W. Bright, Electrostatic charge on hyperbaric chambers, IEEE-IAS Annual Conference Meeting Records, 1979, pp. 192–195.

Katz, B., *Nerve, Muscle, and Synapse*, McGraw-Hill, New York, 1966.

Pethig, R., Dielectric properties of biological materials: biophysical and medical applications, *IEEE Trans. Electr. Insul.*, **EI-19**: 453–474 (1984).

Plonsey, R., *Bioelectric Phenomena*, McGraw-Hill, New York, 1969.

Pohl, H. A., and J. S. Crane, Dielectrophoresis of biological materials, in *Electrostatics and Its Applications*, A. D. Moore, Ed., Wiley, New York, 1973, Chap. 15.

Presman, A. S., *Electromagnetic Fields and Life*, Plenum, New York, 1970.

Reswick, J. B., Development of feedback control prosthetic and orthotic devices, *Adv. Biomed. Eng.*, **2**: 140–218 (1972).

Sances, A., and S. J. Larson, *Electroanesthesia*, Academic, New York, 1975.

Saville, D. A., Electrokinetic effects with small particles, *Annu. Rev. Fluid Mech.*, **9**: 321–37 (1977).

Scott, R. N., Myoelectric control systems, *Adv. Biomed. Eng. Med. Phys.*, **2**: 45–72 (1968).

Shaw, P. J., *Electrophoresis*, Academic, New York, 1969.

Snaddon, R. W., Electrically enhanced collection of respirable aerosols in granular bed filters at low Reynolds numbers, IEEE-IAS Annual Conference Meeting Record, 1982, pp. 1045–1049.

Thomas, H. E., *Handbook of Biomedical Instrumentation and Measurement*, Reston Publishing, Reston, VA, 1974.

Webster, J. G., Ed., *Medical Instrumentation*, Houghton Mifflin, Boston, 1978.

Zimmerman, U., Electric field-mediated fusion and related electrical phenomena, *Biochem. Biophys. Acta A*, **694**: 227–277 (1982).

Communications

Communications systems involve devices for input, for transmission, and for output of information. Electrostatics plays a role in each of these processes, starting with the microphones and phonograph pickups which convert the input to signals, and ending with the headphones and electrostatic speakers at the output. Its role in transmission is less important, but the crystal filter is one example of an electrostatic device which commonly occurs in communications systems. Much of the noise which interferes with transmission is electrostatic in origin and is controlled with electrostatic techniques such as shielding.

Bodle, D. W., et al., *Characterization of the Electrical Environment*, University of Toronto Press, Toronto, 1976.

Ficci, R. F., *Practical Design for Electromagnetic Compatibility*, Hayden, Rochelle Park, NJ, 1971.

Jones, T. B., Lumped parameter electromechanics of electret transducers, *IEEE Trans. Acoust., Speech Signal Process.*, **assp-22**: 141–501 (1974).

Meyer, E., and E.-G. Neumann, Electrostatic transducers, in *Physical and Applied Acoustics*, Academic, New York, 1972, Chap. 8.

Müller, J., Photodiodes for optical communication, *Adv. Electron. Electron Phys.*, **55**: 189–308 (1981).

Olson, H. F., *Acoustical Engineering*, Van Nostrand, Princeton, NJ, 1957.

Electrical Power and Energy Conversion

Electrical power is transmitted at high voltages, so every component of the transmission system must be designed to control the high electrostatic fields encountered. In addition to the lines, cables, and transformers, there are many auxiliary devices, such as capacitors and voltage limiters, which also incorporate electrostatic principles. Electrostatic generators and motors have often been suggested, and have found a few niches, but in general they are not powerful enough for most applications. Electrostatic precipitators are widely used in fossil fuel power plants — and have been for years. At lower voltages, electrostatic effects become important in several electrochemical processes ranging from corrosion of underground conductors to the design of batteries and other devices for direct energy conversion.

Alston, L. L., Ed., *High Voltage Technology,* Oxford University Press, Oxford, 1968.

Bube, R. H., and A. L. Fahrenbruch, Photo-voltaic effect, *Adv. Electron. Electron Phys.,* **56:** 163–217.

Denegri, G. B., et al., Field-enhanced motion of impurity particles in fluid dielectrics under linear conditions, *IEEE Trans. Electr. Insulat.,* **EI-22:** 114–124 (1977).

Hoburg, J. F., Charge density, electric field, and particle charging in electrostatic precipitation with back ionization, *IEEE Trans. Ind. Appl.,* **IA-18:** 666–672 (1982).

Ketani, M. A., *Direct Energy Conversion,* Addison-Wesley, Reading, MA, 1970.

Kuffel, E., and M. Abdullah, *High Voltage Engineering,* Pergamon, Oxford, 1970.

Laghari, J. R., A review of particle-contaminated gas breakdown, *IEEE Trans. Electr. Insul.,* **EI-16:** 388–398 (1981).

Masuda, S., and S. Hosokawa, Performance of two-stage type electrostatic precipitators, *IEEE Trans. Ind. Appl.,* **IA-20:** 709–717 (1984).

Moran, J. H., Electrostatics in the power industry, in *Electrostatics and Its Applications,* A. D. Moore, Ed., Wiley, New York, 1973, Chap. 16.

Peek, F. W., *Dielectric Phenomena in High Voltage Engineering,* McGraw-Hill, New York, 1920.

Soo, S. L., *Direct Energy Conversion,* Prentice-Hall, Englewood Cliffs, NJ, 1968.

Saums, H. L., and W. W. Pendleton, *Materials for Electrical Insulating and Dielectric Functions,* Hayden, Rochelle Park, NJ, 1973.

Von Hippel, A. D., *Dielectric Materials and Applications,* Wiley, New York, 1954.

Electronics

The devices used in solid-state electronics are based on electrostatics, just as the devices used in vacuum tube electronics were. These include resistors, capacitors, and the active devices such as bipolar and field effect transistors (FET), which are based on control of electron motion by electrostatic fields (or the generation of fields by electron motion driven by diffusion and other processes). The fabrication of these devices involves a great deal of additional electrostatics, ranging from clean-room air filters and workstation grounding to ion implantation and electroepitaxial growth. Even in normal operation, electrostatics continues to play a role, as the discharges created by static electricity often cause damage to electronic circuits based on FET technology.

Greason, W. D., and G. S. P. Castle, The effects of electrostatic discharge on microelectronic devices — a review, *IEEE Trans. Ind. Appl.*, **IA-20:** 247–252 (1984).

Hemenway, C. L., et al., *Physical Electronics*, Wiley, New York, 1962.

Horvath, T., and I. Berta, *Static Elimination*, Wiley, New York, 1982.

Jowett, C. E., *Electrostatics in the Electronics Environment*, Wiley, New York, 1976.

Mayer, J. W., et al., *Ion Implantation*, Academic, New York, 1970.

Schlicke, H. M., *Essentials of Dielectromagnetic Engineering*, Wiley, New York, 1961.

Septier, A., Ed., *Applied Charge Particle Optics*, Academic, New York, 1980, Parts A–C.

Sequin, C. H., and M. F. Tompsett, *Charge Transfer Devices*, Academic, New York, 1975.

Spangenberg, K., *Vacuum Tubes*, McGraw-Hill, New York, 1948.

Streetman, B., *Solid State Electronic Devices*, Prentice-Hall, Englewood Cliffs, NJ, 1972.

Thornton, P. R., Electron physics in device microfabrication, *Adv. Electron. Electron Phys.*, **48:** 272–380 (1979).

Van der Ziel, A., *Solid State Physical Electronics*, 3rd ed., Prentice-Hall, Englewood Cliffs, NJ, 1976.

Household and Consumer Goods

In addition to the electronics and communication devices already mentioned, electrostatics plays an important role at home. Most refrigerators and bicycles are painted by an electrostatic spray process which is related to the electrostatic processes used to make sandpaper and velvet flocking for wallpaper. Air is filtered electrostatically, and smoke is detected by changes in leakage currents in a cell. The gas clothes dryer is ignited by a piezoelectric device, and static electricity is removed from the clothes by antistatic chemicals. Watches use liquid crystal displays, and cameras use piezoelectric high voltage generators to fire the flash.

Cobine, J. D., Other electrostatic effects and applications, in *Electrostatics and Its Applications*, A. D. Moore, Ed., Wiley, New York, 1973, Chap. 19.

Eden, H. F., Electrostatic nuisances and hazards, in *Electrostatics and Its Applications*, A. D. Moore, Ed., Wiley, New York, 1973, Chap. 18.

Nowikow, W., The electrostatic behavior of carpets, *J. Electrostat.*, **13:** 249–256 (1982).

Simon, F. N., and Rork, G. D., Ionization type smoke detectors, *Rev. Sci. Inst.*, **49:** 74–80.

Industry (Chemical and Manufacturing)

Electrostatics is most likely to appear in industry during coating operations, in a variety of forms. Paint spraying, electrodeposition, and plasma deposition are all in common use in various industrial contexts. Machining is another area with several applications, including electrospark and electrochemical machining and electrostatic chucks which hold nonmagnetic materials. Separations, especially of fine particles, often take place in electrostatic precipitators, whereas electrohydrodynamic pumping and heat transfer enhancement are useful in some special cases.

Bäkish, R., Ed., *Introduction to Electron Beam Technology*, Wiley, New York, 1962.

Bard, A. J., and L. R. Faulkner, *Electrochemical Methods*, Wiley, New York, 1980.

Bergles, A. E., Techniques to augment heat transfer, in *Handbook of Heat Transfer*, W. M. Roshenow and J. P. Hartnett, Ed., McGraw-Hill, New York, 1973, pp. 10–19 to 10–21.

Birks, J. B., Electrophoretic deposition of insulating materials, *Prog. Dielectr.*, **1**: 271–312 (1959).

Bradley, R. F., and J. F. Hoburg, Electrohydrodynamic augmentation of forced convection heat transfer, *IEEE-IAS Annu. Conf. Meet. Rec.*, 1984, pp. 1146–1150.

Cobine, J. D., Other electrostatic effects and applications, in A. D. Moore, Ed., *Electrostatics and Its Applications*, Wiley, New York, 1973, Chap. 19.

Colver, G. M., and Y. Nakai, Flame stabilization by an electrical discharge and flame visualization of the influence of a semi-insulating wall on the ionic-wind, *IEEE-IAS Annu. Meet. Conf. Rec.*, 1978, pp. 95–104.

Cotroneo, F. A., and G. M. Colver, Electrically augmented pneumatic transport of copper spheres at low particle and duct reynolds numbers, *J. Electrostat.*, **5**: 205–223 (1978).

Dietz, P. W., Heat transfer in bubbling electrofluidized beds, *J. Electrostat.*, **5**: 297–308 (1978).

Eden, H. F., Electrostatic nuisances and hazards, in *Electrostatics and Its Applications*, A. D. Moore, Ed., Wiley, New York, 1973, Chap. 18.

Hoenig, S. A., New applications of electrostatic technology to control of dust, fumes, smokes and aerosols, *IEEE Trans. Ind. Appl.*, **IA-17**: 386–391 (1981).

Hughes, J. F., et al., An experimental electrostatic emulsifier, *IEEE-IAS Annu. Conf. Meet. Rec.*, 1981, pp. 1031–1035.

Jones, T. B., Electrohydrodynamically enhanced heat transfer in liquids — a review, *Adv. Heat Transfer*, **14**: 107–148 (1978).

Kelly, A. J., The electrostatic atomization of hydrocarbons, *Proc. Second Int. Conf. Liquid Atomization and Spray Systems*, Madison, WI., 1982, pp. 57–65.

Loehrke, R. I., and W. J. Day, Performance characteristics of several EHD heat pipe designs, *J. Electrostat.*, **5**: 285–296 (1978).

Masuda, S., and J. Moon, Electrostatic precipitation of carbon soot from diesel engine exhaust, *IEEE Trans. Ind. Appl.*, **IA-19**: 1104–1111 (1983).

Masuda, S., Electric curtain spray booth for powder coating, *IEEE-IAS Annu. Meet. Conf. Rec.*, 1977, pp. 887–892.

Melcher, J. R., K. S. Sachar, and E. P. Warren, Overview of electrostatic devices for control of submicron particles, *Proc. IEEE*, **65**: 1659–1669 (1977).

Miller, E. P., Electrostatic coating, in *Electrostatics and Its Applications*, A. D. Moore, Ed., Wiley, New York, 1973, Chap. 11.

Penney, G. W., and J. Uber, Movement of dust between electrodes in an electrostatically augmented fabric filter, *IEEE-IAS Meet. Conf. Rec.*, 1982, pp. 1067–1070.

Postnikov, S. N., *Electrophysical and Electrochemical Phenomena in Friction, Cutting, and Lubrication*, Van Nostrand Reinhold, New York, 1978.

Raub, E., and K. Müller, *Fundamentals of Metal Deposition*, Elsevier, Amsterdam, 1967.

Robinson, M., Electrostatic precipitation, in *Air Pollution Control*, Part I, Wiley, New York, 1971, pp. 227–336.

Roth, D. G., and A. J. Kelly, Analysis of the disruption of evaporating charged droplets, *IEEE Trans. Ind. Appl.*, **IA-19**: 771–775 (1983).

Roy, A. S., A perspective on electrochemical transport phenomena, *Adv. Heat Transfer*, **12**: 195–282 (1976).

Sickles, J. E., and T. C. Anestos, Electrostatic generation of fine paint droplets, *IEEE Trans. Ind. Appl.*, **IA-15**: 273–276 (1979).

Soo, S., *Fluid Dynamics of Multiphase Systems*, Blaisdell, Waltham, MA, 1967.

Speck, C. E., and M. G. Reynolds, Electrostatic processes affecting attraction of dirt to vehicle bodies during conventional (nonelectrostatic) painting, *IEEE-IAS Annu. Meet. Conf. Rec.*, 1071–1079 (1984).

Van Turnhout, J., et al., Electret filters for high-efficiency and high-flow air cleaning, *IEEE Meet. Conf. Rec.*, 240–248 (1981).

White, H. J., *Industrial Electrostatic Precipitation*, Addison-Wesley, Reading, MA, 1963.

Wilson, J. F., *Practice and Theory of Electrochemical Machinery*, Wiley, New York, 1971.

Information Processing

The role of electrostatics in the electronic chips which make up the heart of the computer has been described before, but the role does not end there. Almost all the peripheral devices, with the exception of magnetic memory, are based in large part on electrostatic effects. Touch pads and capacitance keyboards are two examples of input devices, and output involving CRTs, LCDs, or plasma displays is quite common. Electrostatic printers, like the laser printer, the ink jet printer, or the electrographic printer, are quickly replacing mechanical printers because they offer higher quality and speed, often at lower prices. As robotics becomes more important, there is an increasing need for more varied input and output devices incorporating optical and mechanical transducers, which are often based on electrostatic principles.

Carnahan, R. D., and S. L. Hou, Ink jet technology, *IEEE Trans. Ind. Appl.*, **IA-13**: 95–105 (1977).

Dessauer, J. H., and H. E. Clark, *Xerography and Related Processes*, Focal Press, London, 1965.

Field, D. R., A model to describe electrographic development of resistive one-component toner systems, *IEEE Trans. Ind. Appl.*, **IA-19**: 759–765 (1983).

Hendricks, C. D., Electrostatic imaging, in *Electrostatics and Its Applications*, A. D. Moore, Ed., Wiley, New York, 1973, Chap. 12.

Jackson, R. N., and K. E. Johnson, Gas discharge displays: a critical review, *Adv. Electron. Electron Phys.*, **35**: 191–269 (1974).

Kuhn, L., and R. A. Myers, Ink-jet printing, *Sci. Am.*, 1979, pp. 162–178.

Luxenberg, H. R., and R. L. Kuehn, Eds., *Display Systems Engineering*, McGraw-Hill, New York, 1968.

Ritz, E. F., Recent advances in electron beam deflection, *Adv. Electron. Electron Phys.*, **49**: 299–359 (1979).

Schaffert, R. M., *Electrophotography*, Focal Press, New York, 1966.

Scharfe, M. E., and F. W. Schmidlin, Charged pigment xerography, *Adv. Electron. Electron Phys.*, **38**: 83–147 (1975).

Schein, L. B., Microscopic theory of magnetic brush development, *Photogr. Sci. Eng.*, **19**: 255–265 (1975).

Swatik, D. W., Nonimpact printing, in *Electrostatics and Its Applications*, A. D. Moore, Ed., Wiley, New York, 1973 Chap. 13.

Weimer, P. K., Image sensors for solid state cameras, *Adv. Electron. Electron Phys.*, **37**: 182–263 (1975).

Williams, E. M., *The Physics and Technology of Electrophotographic Processes*, Wiley, New York, 1984.

Instrumentation

Because electrostatic forces are weak, devices based on electrostatics are often used to measure physical quantities without interfering with the process under observations. They are especially useful for measuring forces, pressures, motions, and the flow of gases and liquids. Optical devices such as photodiodes and photocells are quite common, and the related nuclear radiation devices, as far back as the electroscope and the Geiger counter, have always relied on electrostatics to detect the ionization left in the wake of high energy particles. Chemical instrumentation is also replete with electroanalytical devices based on electrochemistry, mass spectrometers, humidity transducers, and ion concentration probes. Instrumentation companies also supply electrostatic generators such as the van der Graaf generator, as well as specialized instruments needed to measure electric fields and charges without disturbing the system under test.

Bard, A. J., and L. R. Faulkner, *Electrochemical Methods*, Wiley, New York, 1980.

Chen, F. F., Electric probes, in *Plasma Diagnostic Techniques*, R. H. Huddlestone and S. L. Leonard, Eds., Academic, New York, 1965, pp. 113–200.

Inhaber, H., Electroecology, in *Physics of the Environment*, Ann Arbor Science, Ann Arbor, MI, Chap. 5, pp. 87–123.

Keithley Corp., *Electrometer Measurements*, Keithley Instruments, Cleveland, OH, 1972.

Lion, K. S., *Instrumentation in Scientific Research*, McGraw-Hill, New York, 1959.

Morrison, R., *Grounding and Shielding Techniques in Instrumentation*, Wiley, New York, 1977.

Rosotti, H., *Chemical Applications of Potentiometry*, Van Nostrand, Princeton, NJ, 1969.

Schwab, A. J., *High Voltage Measurement Techniques*, MIT Press, Cambridge, Mass, 1972.

Secker, P. E., The design of simple instruments for measurement of charge on insulating surfaces, *J. Electrostat.*, 1: 27–36 (1975).

Stimpson, B. P., and C. A. Evans, Electrohydrodynamic ionization mass spectrometry: review of instrumentation, mechanisms and application, *J. Electrostat.*, 5: 411–430 (1978).

Velkoff, H. R., and T. Yuecan, Investigation of electrostatic charge action in liquids for use in detection of cavitation, *IEEE-IAS Meet. Conf. Rec.*, 1984, pp. 1096–1104.

Wilkinson, D. H., *Ionization Chambers and Counters*, Cambridge University Press, Cambridge, 1950.

Mining and Minerals

Prospecting for minerals often involves electrostatic sensors for conductivity and telluric currents. The processing of the ore may involve electrostatic separation of unwanted materials, or electrochemical refining of metals like aluminum. Electrostatics plays an important role in desalination of salt water, and also affects the properties of many minerals, such as clay. It has frequently been blamed for explosions in petroleum and gas transport.

Anderson, J. M., et al., Electrostatic separation of coal macerals, *IEEE Trans. Ind. Appl.*, IA-15: 291–293 (1979).

Klinkenberg, A., and J. L. van der Minne, *Electrostatics in the Petroleum Industry*, Elsevier, Amsterdam, 1958.

Lawver, J. E., and W. P. Dyrenforth, Electrostatic separation, in *Electrostatics and Its Applications*, A. D. Moore, Ed., Wiley, New York, 1973, Chap. 10.

Parkhomenko, E. I., *Electrical Properties of Rocks*, Plenum, New York, 1967.

Ralston, O. C., *Electrostatic Separation of Mixed Granular Solids*, Elsevier, Amsterdam, 1961.

Street, N., Electrokinetics, Circular 263, Illinois State Geological Survey, 1959, pp. 1–18.

Science

For many years, the only major role of electrostatics was to account for the properties of atoms and molecules, which contain positive and negative particles and which interact not only through the Coulomb force but also through the forces induced in polarizible materials. These basic applications are still important, but they have now expanded greatly. Plasma physics, which concerns ionized material, has become especially important owing to the growth of the controlled nuclear fusion effort. The theory of chemical reactions has advanced considerably by considering the intra-atomic electrostatic forces, especially in the field of electrochemistry.

Barreto, E., Electrically produced submicroscopic aerosols, *Aerosol Sci.*, **2**: 219–228 (1971).

Bockris, J. O'M., and A. N. Reddy, *Modern Electrochemistry*, Vols. 1 and 2, Plenum, New York, 1970.

Grivet, P., *Electron Optics*, Pergamon, New York, 1972.

Kelly, A. J., Low charge density electrostatic atomization, *IEEE Trans. Ind. Appl.*, **IA-20**: 267–273 (1984).

Mort, J., and D. M. Pai, *Photoconductivity and Related Phenomena*, Elsevier, Amsterdam, 1976.

Nassar, E., *Fundamentals of Gaseous Ionization and Plasma Electronics*, Wiley, New York, 1971.

Robinson, K. S., R. J. Turnbull, and K. Kim, Electrostatic spraying of liquid insulators, *IEEE Trans. Ind. Appl.*, **IA-16**: 306–317 (1980).

Rose, O. J., and M. Clark, *Plasmas and Controlled Fusion*, Wiley, New York, 1961.

Wehr, M. R., et al., *Physics of the Atom*, Addison-Wesley, Reading, MA, 1978.

Whitby, K. T., and B. Y. H. Liu, The electrical behavior of aerosols, in *Aerosol Science*, C. N. Davies, Ed., Academic, New York, 1966, pp. 59–109.

Woosley, J. P., and R. J. Turnbull, Technique for producing uniform charged drops of cryogenic liquids, *Rev. Sci. Instrum.*, **48** (1977).

APPENDIX A

SOME USEFUL SOLUTIONS OF LAPLACE'S EQUATION

The electrostatic equations have published solutions for hundreds of geometries, but only a few are needed to illustrate the behavior of most electrostatic devices. These solutions are tabulated in the following sections, along with expressions for gradient, divergence, and Laplacian in the three basic coordinate systems used in this book.

Rectangular Geometry

The basic equations are

$$\nabla \Phi = \frac{\partial \Phi}{\partial x} \mathbf{i}_x + \frac{\partial \Phi}{\partial y} \mathbf{i}_y + \frac{\partial \Phi}{\partial z} \mathbf{i}_z \qquad (A.1)$$

$$\nabla \cdot E = \frac{\partial E_x}{\partial x} + \frac{\partial E_y}{\partial y} + \frac{\partial E_z}{\partial z} \qquad (A.2)$$

$$\nabla^2 \Phi = \frac{\partial^2 \Phi}{\partial x^2} + \frac{\partial^2 \Phi}{\partial y^2} + \frac{\partial^2 \Phi}{\partial z^2} \qquad (A.3)$$

The potential distribution for parallel plates (Fig. A.1) is

$$\Phi = K_1 z + K_2 \qquad (A.4)$$

The potential distribution for a hyperbola in a corner (Fig. A.2) is

$$\Phi = K_1 xy + K_2 \qquad (A.5)$$

Cylindrical Geometry

The coordinates (Fig. A.3) are

$$x = r \cos \theta \qquad (A.6a)$$

240

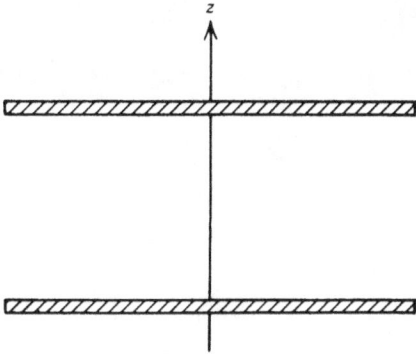

FIGURE A.1. Parallel plates.

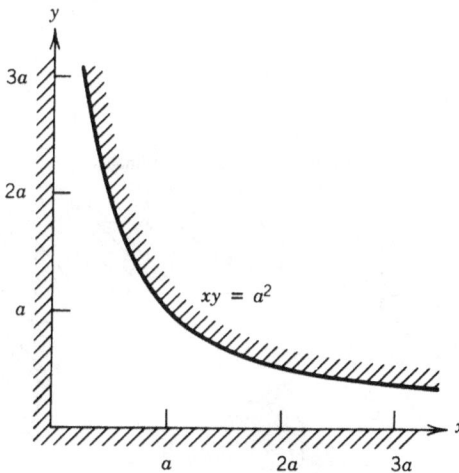

FIGURE A.2. Hyperbola in a corner.

$$y = r \sin \theta \qquad\qquad \text{(A.6b)}$$

$$z = z \qquad\qquad \text{(A.6c)}$$

The basic equations are

$$\nabla \Phi = \frac{\partial \Phi}{\partial r} \mathbf{i}_r + \frac{1}{r} \frac{\partial \Phi}{\partial \theta} \mathbf{i}_\theta + \frac{\partial \Phi}{\partial z} \mathbf{i}_z \qquad\qquad \text{(A.7)}$$

$$\nabla \cdot E = \frac{1}{r} \frac{\partial}{\partial r} (r E_r) + \frac{1}{r} \frac{\partial E_\theta}{\partial \theta} + \frac{\partial E_z}{\partial z} \qquad\qquad \text{(A.8)}$$

$$\nabla^2 \Phi = \frac{1}{r} \frac{\partial}{\partial r} \left(r \frac{\partial \Phi}{\partial r} \right) + \frac{1}{r^2} \frac{\partial^2 \Phi}{\partial \theta} + \frac{\partial^2 \Phi}{\partial z^2} \qquad\qquad \text{(A.9)}$$

The potential between concentric cylinders (Fig. A.4) is

$$\Phi = K_1 \ln r + K_2 \qquad\qquad \text{(A.10)}$$

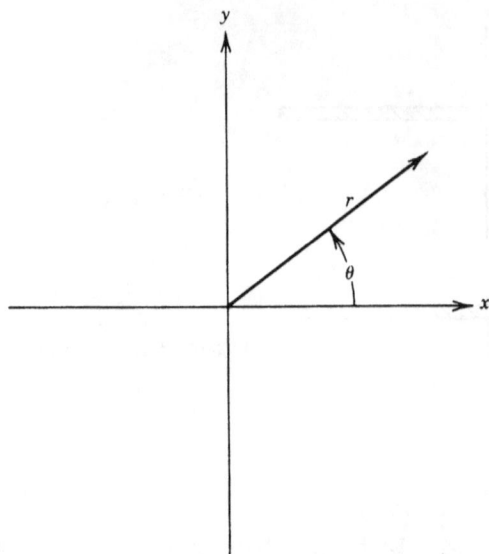

FIGURE A.3. Circular coordinates.

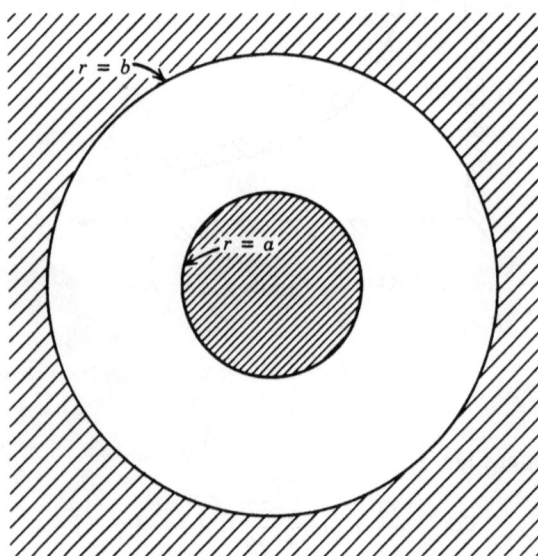

FIGURE A.4. Concentric cylinders.

The potential between infinite wedges (Fig. A.5) is

$$\Phi = K_1\theta + K_2 \qquad (A.11)$$

The potential around a cylinder in an external field (Fig. A.6) is

$$\Phi = K_1 r \cos\theta + \frac{K_2}{r}\cos\theta + K_3 r \sin\theta + \frac{K_4}{r}\sin\theta \qquad (A.12)$$

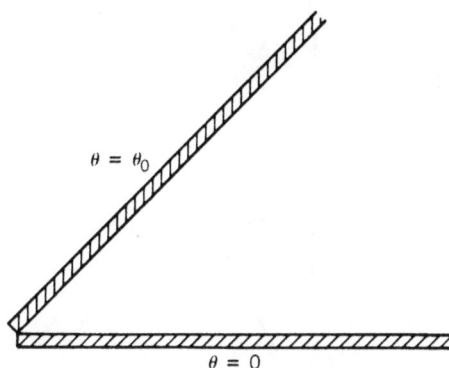

FIGURE A.5. Infinite wedges.

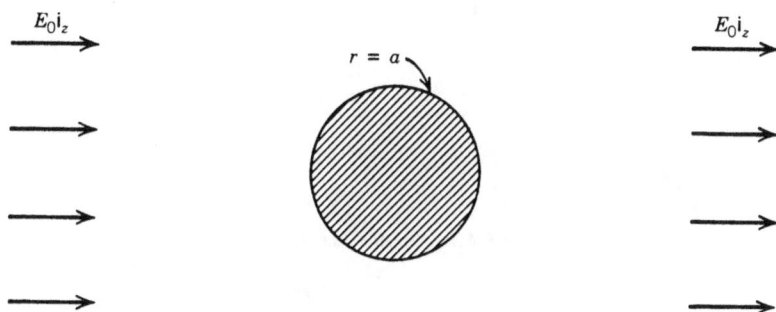

FIGURE A.6. Cylinder in an external field.

Spherical Geometry

The coordinates (Fig. A.7) are

$$x = r \sin \theta \cos \psi \tag{A.13a}$$

$$y = r \sin \theta \sin \psi \tag{A.13b}$$

$$z = r \cos \theta \tag{A.13c}$$

The basic equations are

$$\nabla \Phi = \frac{\partial \Phi}{\partial r} \mathbf{i}_r + \frac{1}{r} \frac{\partial \Phi}{\partial \theta} \mathbf{i}_\theta + \frac{1}{r \sin \theta} \frac{\partial \Phi}{\partial \psi} \mathbf{i}_\psi \tag{A.14}$$

$$\nabla \cdot \mathbf{E} = \frac{1}{r^2} \frac{\partial}{\partial r} (r^2 E_r) + \frac{1}{r \sin \theta} \frac{\partial}{\partial \theta} (\sin \theta E_\theta) + \frac{1}{r \sin \theta} \frac{\partial E_\psi}{\partial \psi} \tag{A.15}$$

$$\nabla^2 \Phi = \frac{1}{r^2} \frac{\partial}{\partial r} \left(r^2 \frac{\partial \Phi}{\partial r} \right) + \frac{1}{r^2 \sin \theta} \frac{\partial}{\partial \theta} \left(\sin \theta \frac{\partial \Phi}{\partial \theta} \right) + \frac{1}{r^2 \sin^2 \theta} \frac{\partial^2 \Phi}{\partial \psi^2} \tag{A.16}$$

The potential between concentric spheres (Fig. A.8) is

$$\Phi = \frac{K_1}{r} + K_2 \tag{A.17}$$

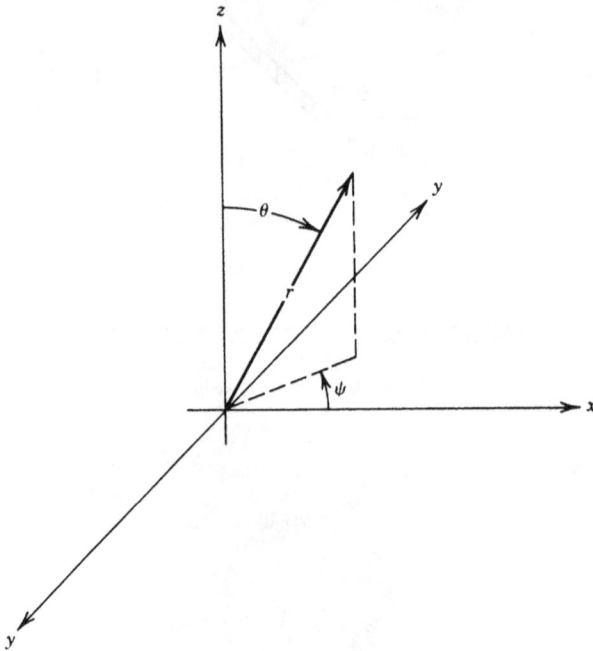

FIGURE A.7. Spherical coordinates.

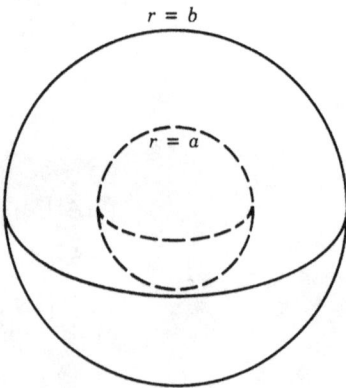

FIGURE A.8. Concentric spheres.

The potential between infinite cones (Fig. A.9) is

$$\Phi = K_1 \ln \cot \frac{\theta}{2} + K_2 \qquad (A.18)$$

The potential around a sphere in uniform field (Fig. A.10) is

$$\Phi = K_1 r \cos \theta + \frac{K_2}{r^2} \cos \theta + K_3 r \sin \theta + \frac{K_4}{r^2} \sin \theta \qquad (A.19)$$

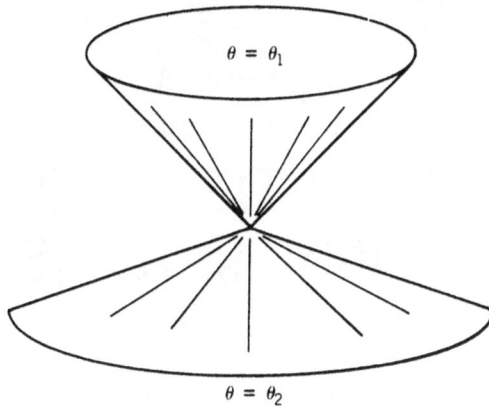

FIGURE A.9.   Infinite cones.

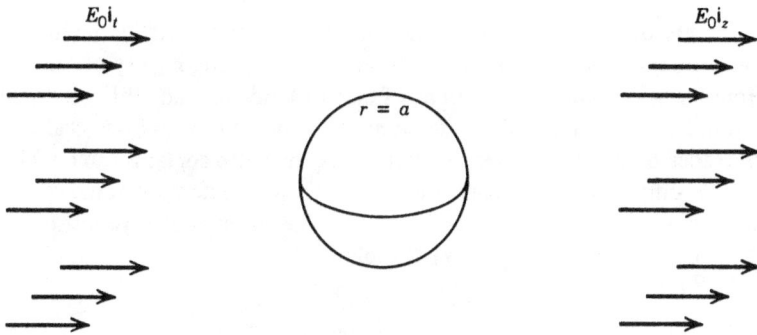

FIGURE A.10.   Sphere in a uniform field.

BIBLIOGRAPHY

Durand, E., *Electrostatique,* Vols. 1–3, Masson, Paris, 1960.

Moon, P., and D. E. Spencer, *Field Theory Handbook,* Springer-Verlag, Berlin, 1961.

Moon, P., and D. E. Spencer, *Field Theory for Engineers,* Van Nostrand, Princeton, NJ, 1961.

Smythe, W. R., *Static and Dynamic Electricity,* McGraw-Hill, New York, 1950.

===

SI UNITS AND CONVERSIONS

Although this book uses SI units, the literature of electrostatics includes many nonstandard units. Part of this variety comes from the long history of the field and part from the widespread applications of electrostatics in many different areas. As a result, much of the worthwhile literature uses other systems of physical units. To aid the reader in using the older literature, this appendix contains three tables covering the SI units, the SI prefixes, and the conversion factors for some of the more common units which appear in the literature. The SI units of several fundamental constants are also included in a fourth table.

TABLE B.1. SI Units

Quantity	Name	Common Unit	Fundamental Unit
Acceleration		m/s^2	$m \cdot s^{-2}$
Admittance	siemans	S	$m^{-2} \cdot kg^{-1} \cdot s^3 \cdot A^2$
Amount of substance	mole	mol	mol
Angle (plane)	radian	rad	rad
(solid)	steradian	sr	sr
Angular velocity		rad/s	$rad \cdot s^{-1}$
Angular acceleration		rad/s^2	$rad \cdot s^{-2}$
Angular momentum		$kg \cdot rad/s$	$kg \cdot rad \cdot s^{-1}$
Capacitance	farad	F	$m^{-2} \cdot kg^{-1} \cdot s^4 \cdot A^2$
Charge (electric)	coulomb	C	$s \cdot A$
Charge density		C/m^3	$m^{-3} \cdot s \cdot A$
Conductance (electric)	siemans	S	$m^{-2} \cdot kg^{-1} \cdot s^3 \cdot A^2$
Conductivity (electric)		S/m	$m^{-3} \cdot kg^{-1} \cdot s^3 \cdot A^2$
(thermal)		$W/(m \cdot K)$	$m \cdot kg \cdot s^{-3} \cdot K^{-1}$
Current (electric)	ampere	A	A
Current density		A/m^2	$m^{-2} \cdot A$
Density (mass)		kg/m^3	$kg \cdot m^{-3}$
(charge)		C/m^3	$m^{-3} \cdot s \cdot A$
(surface charge)		C/m^2	$m^{-2} \cdot s \cdot A$
(line charge)		C/m	$m^{-1} \cdot s \cdot A$
Dielectric constant (dimensionless)			
Diffusion constant		m^2/s	$m^2 \cdot s^{-1}$
Dipole moment (electric)		$C \cdot m$	$m \cdot s \cdot A$
Electric field strength		V/m	$m \cdot kg \cdot s^{-3} \cdot A^{-1}$
Electric flux density (displacement)		C/m^2	$m^{-2} \cdot s \cdot A$
Electromotive force	volt	V	$m^2 \cdot kg \cdot s^{-3} \cdot A^{-1}$
Energy	joule	J	$m^2 \cdot kg \cdot s^{-2}$
Energy density		J/m^3	$m^{-1} \cdot kg \cdot s^{-2}$
Frequency	hertz	Hz	s^{-1}
Frequency (angular)	radian	rad/s	$s^{-1} \cdot rad$
Force	newton	N	$m \cdot kg \cdot s^{-2}$
Force density		N/m^3	$m^{-2} \cdot kg \cdot s^{-2}$
Heat	joule	J	$m^2 \cdot kg \cdot s^{-2}$
Heat current density		W/m^2	$kg \cdot s^{-3}$
Impedance	ohm	Ω	$m^2 \cdot kg \cdot s^{-3} \cdot A^{-2}$
Inductance	henry	H	$m^2 \cdot kg \cdot s^{-2} \cdot A^{-2}$
Length	meter	m	m
Magnetic field strength		A/m	$m^{-1} \cdot A$
Magnetic flux	weber	Wb	$m^2 \cdot kg \cdot s^{-2} \cdot A^{-1}$
Magnetic flux density	tesla	T	$kg \cdot s^{-2} \cdot A^{-1}$
Mass	kilogram	kg	kg
Mobility		$m^2/V\text{-}s$	$kg^{-1} \cdot s^2 \cdot A$
Momentum		$kg \cdot m/s$	$m \cdot kg \cdot s^{-1}$
Moment of force		$N \cdot m$	$m^2 \cdot kg \cdot s^{-2}$

TABLE B.1. *(continued)*

Quantity	Name	Common Unit	Fundamental Unit
Number density		number/m^3	m^{-3}
Particle flux		number/m^2-s	m$^{-2} \cdot$ s^{-1}
Permeability		H/m	m $\cdot$ kg $\cdot$ s$^{-2} \cdot$ A^{-2}
Permittivity		F/m	m$^{-3} \cdot$ kg$^{-1} \cdot$ s$^4 \cdot$ A^2
Polarization		C/m^2	m$^{-2} \cdot$ s $\cdot$ A
Polarizability (material)		F/m	m$^{-3} \cdot$ kg$^{-1} \cdot$ s$^4 \cdot$ A^2
Potential (electric)	volt	V	m$^2 \cdot$ kg $\cdot$ s$^{-3} \cdot$ A^{-1}
Power	watt	W	m$^2 \cdot$ kg $\cdot$ s^{-3}
Pressure	pascal	Pa	m$^{-1} \cdot$ kg $\cdot$ s^{-2}
Resistance (electric)	ohm	Ω	m$^2 \cdot$ kg $\cdot$ s$^{-3} \cdot$ A^{-2}
Stress	pascal	Pa	m$^{-1} \cdot$ kg $\cdot$ s^{-2}
Surface tension		N/m	kg $\cdot$ s^{-2}
Susceptibility (electric)			dimensionless
Temperature (absolute)	kelvin	K	K
(conventional)	degree Celsius	°C	K
Time	second	s	s
Torque		N $\cdot$ m	m$^2 \cdot$ kg $\cdot$ s^{-2}
Velocity		m/s	m $\cdot$ s^{-1}
Viscosity (dynamic)		Pa $\cdot$ s	m$^{-1} \cdot$ kg $\cdot$ s^{-1}
(kinematic)		m^2/s	m$^2 \cdot$ s^{-1}
Voltage	volt	V	m$^2 \cdot$ kg $\cdot$ s$^{-3} \cdot$ A^{-1}

TABLE B.2. **SI Prefixes**

Prefix	Symbol	Factor
exa	E	10^{18}
peta	P	10^{15}
tera	T	10^{12}
giga	G	10^9
mega	M	10^6
kilo	k	10^3
hecto	h	10^2
deka	da	10^1
deci	d	10^{-1}
centi	c	10^{-2}
milli	m	10^{-3}
micro	μ	10^{-6}
nano	n	10^{-9}
pico	p	10^{-12}
femto	f	10^{-15}
atto	a	10^{-18}

TABLE B.3. SI Equivalents

Old Unit	SI Equivalent
angstrom	1.000×10^{-10} m
atmosphere	1.013×10^5 Pa
bar	1.000×10^5 Pa
Btu	1.055×10^3 J
Btu $\cdot$ in/ft$^2 \cdot$ s $\cdot$ °F (thermal conductivity)	5.189×10^2 W/m $\cdot$ K
calorie	4.184 J
calorie (kilogram)	4.184×10^3 J
cm Hg (0°C)	1.333×10^3 Pa
cm H$_2$O (4°C)	9.806×10^1 Pa
centipoise	1.000×10^{-3} Pa $\cdot$ s
centistoke	1.000×10^{-6} m^2/s
circular mil	5.067×10^{-10} m^2
cubic feet/min (cfm)	4.719×10^{-4} m^3/s
day	8.640×10^4 s
degree (angle)	1.745×10^{-2} rad
dyne	1.000×10^{-5} N
dyne/cm^2	1.000×10^{-1} Pa
electron volt	1.602×10^{-19} J
electrostatic unit (esu)	3.336×10^{-10} C
erg	1.000×10^{-7} J
faraday	9.649×10^4 C
fluid ounce (U.S.)	2.957×10^{-5} m^3
foot	3.048×10^{-1} m
foot of water	2.989×10^3 Pa
foot-pound (energy)	1.356 J
foot-pound (torque)	1.356 N-m
gallon (U.S. liquid)	3.785×10^{-3} m^3
gallon per minute (gpm)	6.309×10^{-5} m^3/s
gauss	1.000×10^{-4} T
horsepower	7.460×10^2 W
inch	2.540×10^{-2} m
inch of mercury (60°F)	3.377×10^3 Pa
inch of water (60°F)	2.488×10^2 Pa
inch-ounce (torque)	7.062×10^{-3} N-m
liter	1.000×10^{-3} m^3
kilocalorie	4.184×10^3 J
kilogram force (kgf)	9.807 N
kgf/cm^2	9.807×10^4 Pa
kilometer/hr	2.778×10^{-1} m/s
kilowatt-hour	3.600×10^6 J
lbf (pound force)	4.448 N
lbm (pound mass)	4.536×10^{-1} kg
maxwell	1.000×10^{-8} Wb
mil	2.540×10^{-5} m

TABLE B.3. *(continued)*

Old Unit	SI Equivalent
mile (U.S. statute)	1.609×10^3 m
mile/hour	4.470×10^{-1} m/s
millibar	1.000×10^2 Pa
mm Hg (0°C)	1.333×10^2 Pa
oersted	7.958×10^1 A/m
ounce force	2.780×10^{-1} N
pica (printers)	4.218×10^{-3} m
point (printers)	3.515×10^{-4} m
poise	1.000×10^{-1} Pa · s
pound force (lbf)	4.448 N
pound mass (lbm)	4.536×10^{-1} kg
psi	6.895×10^3 Pa
statampere	3.336×10^{-10} A
statcoulomb	3.336×10^{-10} C
statfarad	1.113×10^{-12} F
statvolt	2.998×10^2 V
stoke	1.000×10^{-4} m²/s
torr	1.333×10^2 Pa
volt per mil	3.937×10^4 V/m
watt-hour	3.600×10^3 J
year	3.154×10^7 s

TABLE B.4. Fundamental Constants

Avagadro constant	6.022×10^{26} kmol^{-1}
Boltzmann constant	1.381×10^{-23} J/K
Electron charge	1.602×10^{-19} C
Electron rest mass	9.110×10^{-31} kg
Faraday constant	9.649×10^7 C/kmol
Permittivity of vacuum	8.854×10^{-12} F/m
Planck constant	6.626×10^{-34} J · s
Proton rest mass	1.673×10^{-27} kg
Speed of light in vacuum	2.998×10^8 m/s

BIBLIOGRAPHY

Goldman, The metric system: its status and future, *IEEE Spectrum*, April 1981, pp. 6–63.

Mechtly, E. A., *The International System of Units*, NASA, Washington, Publication NASA SP-7012, 1973.

Weast, R. C., Ed., *Handbook of Chemistry and Physics*, CRC Press, Cleveland, OH, 1973.

INDEX

www.ingramcontent.com/pod-product-compliance
Lightning Source LLC
Chambersburg PA
CBHW021033210326
41598CB00016B/1008